SpringerBriefs present concise summaries of cutting-edge research and practical applications across a wide spectrum of fields. Featuring compact volumes of 50 to 125 pages, the series covers a range of content from professional to academic.

Typical publications can be:

- A timely report of state-of-the art methods
- An introduction to or a manual for the application of mathematical or computer techniques
- A bridge between new research results, as published in journal articles
- A snapshot of a hot or emerging topic
- An in-depth case study
- A presentation of core concepts that students must understand in order to make independent contributions

SpringerBriefs are characterized by fast, global electronic dissemination, standard publishing contracts, standardized manuscript preparation and formatting guidelines, and expedited production schedules.

On the one hand, **SpringerBriefs in Applied Sciences and Technology** are devoted to the publication of fundamentals and applications within the different classical engineering disciplines as well as in interdisciplinary fields that recently emerged between these areas. On the other hand, as the boundary separating fundamental research and applied technology is more and more dissolving, this series is particularly open to trans-disciplinary topics between fundamental science and engineering.

Indexed by EI-Compendex, SCOPUS and Springerlink.

More information about this series at https://link.springer.com/bookseries/8884

Mohammad Reza Barzegar-Bafrooei ·
Jamal Dehghani-Ashkezari ·
Asghar Akbari Foroud · Hassan Haes Alhelou
Editors

Fault Current Limiters

Concepts and Applications

 Springer

Editors
Mohammad Reza Barzegar-Bafrooei
Department of Electrical Engineering
Ardakan University
Ardakan, Iran

Asghar Akbari Foroud
Electrical and Computer Engineering
Faculty
Semnan University
Semnan, Iran

Jamal Dehghani-Ashkezari
Yazd Electrical Distribution Company
Yazd, Iran

Hassan Haes Alhelou
University College Dublin
Dublin, Ireland

ISSN 2191-530X   ISSN 2191-5318   (electronic)
SpringerBriefs in Applied Sciences and Technology
ISBN 978-981-16-6650-6   ISBN 978-981-16-6651-3   (eBook)
https://doi.org/10.1007/978-981-16-6651-3

This Springer imprint is published by the registered company Springer Nature Singapore Pte Ltd.
The registered company address is: 152 Beach Road, #21-01/04 Gateway East, Singapore 189721,
Singapore

Contents

List of Figures

List of Tables

Chapter 1
Introduction

**Mohammad Reza Barzegar-Bafrooei, Jamal Dehghani-Ashkezari,
Asghar Akbari Foroud, and Hassan Haes Alhelou**

1.1 Introduction

At this time, the management of fault current in the modern transmission and distribution (T&D) systems is considered as a challenging issue within researchers and government and private electric utilities. Since some of the existing systems and most of the future systems would face with high fault current level. The potential world economic growth in recent decades and consequently the demand for electric energy are the main reasons for this problem. It is evident that the electricity utilities should be continuously upgraded their systems to handle the demand. The construction of new power plants, the interconnection of the networks, the application of distributed generations (DGs), and the utilization of parallel transmission lines and flexible AC transmission system (FACTS) are some of the upgrading programs. In addition to demand response, these activities pursue the important power system facets such as higher efficiency and reliability, longer service availability, and dwindling power loss, maintenance and development costs, environmental effects, and dependency on fossil fuels.

M. R. Barzegar-Bafrooei (✉)
Department of Electrical Engineering, Ardakan University, Ardakan, Iran
e-mail: m.barzegar@ardakan.ac.ir

J. Dehghani-Ashkezari
Yazd Electrical Distribution Company, 8916794637 Yazd, Iran

A. Akbari Foroud
Electrical, Computer Engineering Faculty, Semnan University, Semnan, Iran
e-mail: aakbari@semnan.ac.ir

H. Haes Alhelou
The Department of Electrical and Computer Engineering, Isfahan University of Technology, Isfahan, Iran

M. R. Barzegar-Bafrooei et al. (eds.), *Fault Current Limiters*,
SpringerBriefs in Applied Sciences and Technology,
https://doi.org/10.1007/978-981-16-6651-3_1

In spite of beneficiaries, the aforementioned activities may push the fault current level beyond the short circuit capacity of equipment that would increase the risk of power system damage. The elimination of fault occurrence in the power system is unlikely, but their catastrophic impacts can be decreased. Among different types of fault appearing in the power system, short circuit fault is one of the most important and destructive contingencies. Synchronous generators, large induction motors, and DGs mainly feed fault current [1]. Synchronous generators have a significant effect on the magnitude of fault current in steady-state conditions. Whereas, the major effect of a large induction motor is related to the transient component of fault current. The contribution of DGs in fault current may be low or high depending on DG type. For example, inverter-based DGs such as photovoltaic systems have a minor contribution. Generally, the fault current can be as high as more than 10 times the maximum nominal current. It even may exceed further in extreme cases [1]. Fault current problem has been increasingly important over the years as the power system becomes more complicated due to load growth. The major problems with enormous fault current are [1–4]:

- The peak of short-circuit current causes high electromagnetic forces and mechanical dynamic stress on equipment beyond their endurable;
- Thermal stress on equipment, especially at the fault location. In case of an extreme fault, conductors and insulation may melt and cause the explosion and fire of equipment;
- Circuit breakers (CBs) experience more severe conditions during fault current interruption or even fail in their duty. Failure of CB may cause significant damages;
- Insulation breakdown may happen due to appeared large surge waves caused by fault current;
- Current transformers may encounter with saturation problem and consequently the protection system makes the wrong decision. The remaining fault for a longer time in the system, outage, loss of synchronization, overvoltage and under-voltage transients, and putting in danger the integrity and stability of the power system are the possible outcomes of this wrong decision;
- The demand for reactive power increases. Since the series impedance of the transformer and transmission line is mostly inductive;
- The magnitude of voltage sag in un-faulted buses increases and affects the power quality indices and the performance of loads, especially voltage-sensitive loads.

One of the most straightforward ways to cope with the high fault current and its adverse impacts on the power system is to replace or upgrade equipment with higher short circuit capacity. However, it is impossible to carry out in all locations due to economic and technical problems. For instance, Table 1.1 lists the forecasted fault current in three different years for 500 kV china's grid [5]. The installed CBs are able to interrupt fault current successfully up to 63 kA. It is difficult to interrupt fault current by CB for fault current above this value. Replacing the CB is not always easy. Sometimes, it may not be available in the market, or it may be produced by special manufacturers. Even, it may be expensive or needs more space to install.

Table 1.1 Magnitude of fault current in 500 kV china's grid [5]

Year	Three-phase fault		Single-phase fault		Number of 500kV substation with high fault current problem
	Max	Average	Max	Average	
2010	62	45.7	61.8	41.6	0
2015	70.9	47.9	72.9	43	13
2020	78.7	52.7	72.5	48	17

In some cases, sequential tripping is also used in the substation to handle fault current without replacement. This strategy can be used in the case where the new source (or sources) is added to the substation, and fault current level becomes more than existing CBs capability. In the method, the implemented CBs for the new source should be tripped before other existing CBs. In fact, the new source is disconnected first and then other underrated CBs operate. It is obvious that this method does not reduce the total fault current. It only eliminates the contribution of the new source in fault by changing the topology of the system during fault conditions and provides safe conditions for the operation of CBs. However, the implementation of sequential tripping is complex and may increase overstress on equipment due to reaming fault for a longer period. Moreover, it is applicable for the substations that are supplied from multiple sources.

Ordinary, fault current limitation techniques are the measures that can reduce fault current in the modern power system. Table 1.2 describes the general classification for fault current limitation techniques. Functionally, from the standpoint of how to apply, fault current limitation techniques can be categorized into topological and equipment-based approaches. Sequential tripping, transferring to higher voltage level, and the reduction of the degree of meshing are the topological-based approaches. High short circuit impedance transformer, current limiting reactors (CLRs), stand-alone high voltage fuse, fuse-like explosive devices (also known as commutation fuse-based limiter), and fault current limiters (FCLs) are the viable options for equipment-based approaches. The terms "active" and "passive" are also used in some literature to classify fault current limitation techniques. The topology of the system does not change before and after fault occurrence in "passive" approaches. The term "active"

Table 1.2 Classification of fault current limitation techniques [1, 3]

Topological-based approaches	• Sequential tripping • Transferring to the higher voltage level • The reduction of the degree of meshing	Passive
Equipment-based approaches	• The high short circuit impedance transformer • Current limiting reactors	Active
	• High voltage fuse • Fuse-like explosive devices • Fault current limiters	

refers to the approaches that are "invisible" from the power system during normal conditions and are only activated during fault conditions [1].

Increasing the voltage level is not the preferred option in many cases due to the significant investment, especially in high voltage. The reduction of the degree of meshing can be achieved by the reconfiguration measures such as splitting into sub-grids or bus-bar splitting. Contributing of different system sections in fault current can be eliminated by dividing a grid into sub-grids (with the same voltage level) or splitting bus-bars by bus-tie CBs that result in reducing the overall fault current level. Although these activities can be applied in fast-growing areas, but, they may reduce the reliability and stiffness of the system or impose the cost of required switchgear. Since the number of sources connected to the buses reduces and dispatching power sacrifices [6].

The transformer with high short circuit impedance and CLR add mostly reactive impedance in the system. CLR installation is preferred in the existing system. Embedding the system by the transformer with high short circuit impedance is interested in a new power system as an upgrade program of exchanging transformers is not a cost-effective way in comparison with CLR installation. As an important drawback, the provided impedance by these methods is permanent and causes voltage drop and power loss during normal operation.

As a simple, cheap, small size, reliable, and quick reaction protective device, fuse is widely used in the distribution system. It operates in less than half a cycle without any fault detection system. However, it is a single-use device and should be manually replaced after each operation, and thereby, a longer service unavailability happens in the system. Furthermore, the fuse is not available for high-rated voltage and current due to technical problems. The rated current restriction can be solved by the utilization of fuse-like explosive devices such as Is-Limiter [7], but the rated voltage problem remains. Indeed, fuse and fuse-like explosive devices are applicable up to medium voltage (MV) level.

In contrast to the aforementioned measures, a new and effective concept that has attracted attention in recent decades is the application of FCL. It is connected in series with the power system and can resolve the concerns raised by power system expansion. Embedding the power system with FCL offers negligible voltage drop during normal operation, rapid and reliable current limitation, and spontaneous recovery. In the existing system, FCL allows the equipment to remain in service without replacement or upgrading plan postpones. Moreover, it allows using the device with lower short circuit capacity and subsequently lower prices in newly planned systems. The utilities' need for FCL and its technical and economic verifications can be confirmed through the government and private companies' investments, technical reports, and papers. As per background, the future T&D systems will be equipped with FCL, and it is needed to be verified all aspects before its application. This book provides the research activities related to this device. Characteristics, technologies, topologies, working principles, applications, and their interaction with the power system are described in the next chapter.

1.2 FCL Concept

FCL is a device or apparatus that would take the preferred option to provide the safeguard against high fault current level in the future power system. According to the definition of IEEE STD C37.02, FCL is equipment with the ability to moderate the first peak and subsequent root mean square (RMS) of fault current. This activity is provided through emerging the resistive and/or reactive impedance during the current limiting mode [8]. The FCL may consist of discrete, functionally integrated, and spatially separated equipment. Figure 1.1 shows the role of FCL in a simple power system. With the expansion of the power system and technical advancements of renewable energy generation, a large number of new power plants and DGs are expected to be connected in power grids. The parallel connection of sources potentially reduces the equivalent impedance of the source. It leads to an increase in the fault current level. The system can be modified by the installation of FCL and subsequently, the concerns associated with fault current are eliminated.

Fault current characteristics in the presence of FCL can be also assessed by analytical studies. Figure 1.2 illustrates the equivalent circuit of a simple power system, shown in Fig. 1.1b. Once a short circuit occurs in the system, the current starts to rise with the definite rate, which depends on system parameters including source resistance (R_S), source inductance (L_S), load angle (δ), and voltage angle at fault inception time (α). In the case of installation of FCL, the parameters of FCL including two constant resistive (R_{FCL}) and inductive (L_{FCL}) parts should be also added in

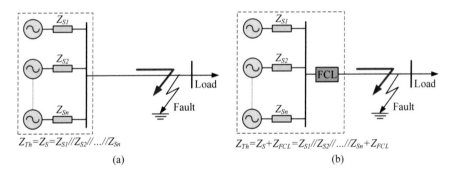

$$Z_{Th}=Z_S=Z_{S1}//Z_{S2}//...//Z_{Sn}$$
(a)

$$Z_{Th}=Z_S+Z_{FCL}=Z_{S1}//Z_{S2}//...//Z_{Sn}+Z_{FCL}$$
(b)

Fig. 1.1 Expression of FCL role in a simple power system; **a** without FCL, **b** with FCL

Fig. 1.2 Equivalent circuit of the simple power system with FCL

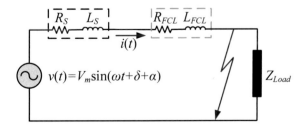

the simplest model. The following equation describes the behavior of fault current considering FCL impedance [9]:

$$(L_S + L_{FCL})\frac{di(t)}{dt} + (R_S + R_{FCL})i(t) = V_m \sin(\omega t + \delta + \alpha) \qquad (1.1)$$

The instantaneous fault current can be calculated after solving the above first-order differential as follows:

$$i(t) = i_{AC}(t) + i_{DC}(t) \Rightarrow i(t) = \frac{V_m}{Z}\left[\sin(\omega t + \alpha + \delta - \varphi) - e^{-t/\tau}\sin(\alpha + \delta - \varphi)\right]$$

$$(1.2)$$

where:

$$Z = \sqrt{(R_S + R_{FCL})^2 + \omega^2(L_S + L_{FCL})^2} \qquad (1.3)$$

$$\varphi = \tan^{-1}\left(\frac{\omega(L_S + L_{FCL})}{R_S + R_{FCL}}\right) \qquad (1.4)$$

$$\tau = \frac{L_S + L_{FCL}}{R_S + R_{FCL}} \qquad (1.5)$$

As seen, FCL installation increases the provided impedance Z in the fault loop, which helps further reduction of the amplitude of AC and DC components of fault current. The resistive part of FCL impedance reduces the time constant (τ), whereas the inductive part increases it. Indeed, the type of the provided impedance by FCL affects the damping rate of the transient component of fault current. It is rapidly damped in the presence of resistive-based FCLs. Another parameter affected by the type of FCL impedance is the power factor during fault conditions (φ). It is clear that resistive type FCL improves the power factor. It is well known that the X/R ratio of the real system is high and the equivalent impedance of the power system during fault conditions is mainly inductive. It implies that the effect of inductive type FCL on time constant and power factor may be low [9–10].

It should be noted that the fault analysis in real power system is more complicated than the simple circuit discussed earlier. The above explanations only address the role of FCL in the power system and cannot verify all issues associated with the behavior of fault current in the presence of FCL. For example, the constant impedance was considered for FCL in Eqs. (1.1)–(1.5), whereas the provided impedance by FCL in the current limiting mode is a function of different factors such as topology and current. Moreover, fault resistance and the exact model of DGs and synchronous generators (especially, sub-transient and transient inductances) should be taken into account.

1.3 FCL Characteristics

Figure 1.3 graphically shows the typical working principle of an FCL and its impacts on fault current. In normal operation, it provides the conducting path with the low impedance to avoid voltage drop and power losses. When a fault occurs in the power system, it should react and limit fault current by inserting a high impedance. As soon as the fault is cleared, FCL would enter in recovery mode. The recovery time should be short and FCL returns in service to carry line current or limitation function duty. Regarding the developed structure for FCL, the insertion impedance may be either constant during the current limiting mode or variable. Furthermore, resultant impedance by FCL operation may depend upon fault current magnitude. Generally, the following requirements are expected from an ideal FCL [11]:

- FCL should virtually disappear during normal conditions, meaning that the voltage drop across FCL and power losses should be near to zero. Moreover, current and voltage waveforms distortion should be also minimized (low harmonic injection);
- It should be able to detect and limit the fault current as soon as possible (fast response time). The fault current limitation duty prior to the first peak is desirable;
- Both DC and AC components of fault current should be effectively limited by FCL without dependency on systems parameters such as fault type and fault resistance;
- Reasonable lifetime;
- Continuous operation, meaning that the current limiting mode should be continued until fault clearance time or allowable time considered by the manufacturer;
- Fast recovery process without human intervention (automatic recovery), especially when it is used in reclosing scheme;
- When sequential fault events in a short time interval happen in the power system, the successful frequent operations are favored for FCL;
- No restriction for development in different ratings;
- High reliability and the low failure rate, meaning that FCL remains in current limiting mode even in case of failure of its primary mechanism;
- The wrong trip due to starting current of induction motors, inrush current of capacitor banks and transformers, and transient currents caused by single phase to ground faults should be avoided;
- There is no concern associated with build-up overvoltage across the FCL;
- Coordination with other conventional protection devices including CBs and relays;
- Compact design, for transportation and likely indoor substation and urban area installation;
- Low environmental impacts.

Additionally, the economic benefits of applying an FCL should be reasonable in comparison with other fault current limitation techniques. The costs associated with an FCL includes research and development (R&D) and long-term costs. At now, R&D costs of an FCL are high due to low practical experience and unforeseen challenges during development. The designed FCL is firstly examined in laboratories

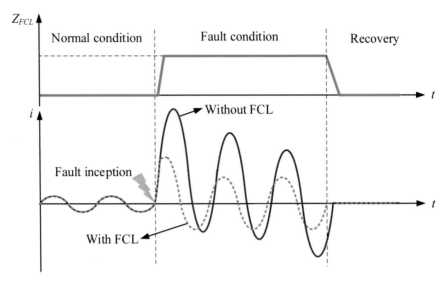

Fig. 1.3 The role of FCL in power system

to avoid unwanted failures. Then the prototype is installed in the specified system to obtain long-term performance in real conditions. The above procedure is time-consuming and costly. However, it is expected that this cost would decrease in the future regarding the obtained experience from the prototype. Long-term costs are also related to power loss and maintenance costs during the useful lifetime. These costs are fixed and should be at a satisfactory level.

In reality, it is impossible to satisfy all of the above-mentioned characteristics to construct an FCL. The ongoing proposed structures for FCL only meet some requirements. Generally, the compromises and certain trade-off considering key aspects should be done to develop an FCL. These parameters may be different depending on the applications of FCL. As an example, in the reclosing scheme, the recovery time and repeated operations are vital. Typically, the current limiting percentage, fast response time, adequate lifetime, the price of FCL, and recovery process are the key aspects that should be considered in the development of an FCL [12]. According to IEEE STD C37.302 [8], developers are conducted to focus on issues including electrical parameters, physical and operational attributes, environmental aspects, safety, and lifetime. The benefits offered by the installation of an ideal FCL in the power system can be summarized are follows [1, 11]:

- The stresses imposed on power system equipment moderates, and so, aging decreases;
- It does not need to replace power equipment with higher short circuit ratings;
- The upgrading plan can be postponed;
- The improvement in transient stability can be achieved;

- The magnitude of voltage sag reduces. It implies that the voltage-sensitive loads connected to the faulted bus and near buses are less affected. The enhancement of the stability of induction motors is concluded;
- The voltage stability can be improved due to avoiding rapid voltage reduction;
- The problems related to CT saturation reduce;
- The power system equipment with lower short circuit ratings can be chosen in the new planed system, meaning that the size and weight of equipment and investment decrease.

Chapter 2
The Classification of FCL

Mohammad Reza Barzegar-Bafrooei, Jamal Dehghani-Ashkezari, Asghar Akbari Foroud, and Hassan Haes Alhelou

In the attempts to develop an FCL that satisfies the requirements of the power system, a large number of FCL technologies have been proposed by researchers and companies. Proposing novel topologies or improving the previous structures based on depth computer analysis, laboratory experimental tests, and live-grid deployment are the consequence of these activities. Several categories have been proposed in papers, standard, and technical reports to classify FCL structures [1, 2, 8, 11, 12]. IEEE standard divides FCLs into two sets consisting of Type A and Type B. The FCL capability to the current interruption is the distinct property between Type A and Type B. The former is only capable to limit fault current, whereas the latter is able to interrupt fault current as well as current limiting. In other words, Type B plays the role of FCL and CB. In a more detailed classification, they individually have two subsets namely Type A1, Type A2, Type B1, and Type B2. The insertion impedance by Type A1 and Type B1 into the system is linear during fault conditions, whereas the nonlinear impedance is provided by Type A2 and Type B2 FCLs. This issue can be graphically illustrated with the help of Fig. 2.1. It should be noted that the experimental waveforms have more distortion.

M. R. Barzegar-Bafrooei
Department of Electrical Engineering, Ardakan University, Ardakan, Iran
e-mail: m.barzegar@ardakan.ac.ir

J. Dehghani-Ashkezari
Yazd Electrical Distribution Company, 8916794637 Yazd, Iran

A. Akbari Foroud
Electrical, Computer Engineering Faculty, Semnan University, Semnan, Iran
e-mail: aakbari@semnan.ac.ir

H. Haes Alhelou (✉)
The Department of Electrical and Computer Engineering, Isfahan University of Technology, Isfahan, Iran

© The Author(s), under exclusive license to Springer Nature Singapore Pte Ltd. 2022   11
M. R. Barzegar-Bafrooei et al. (eds.), *Fault Current Limiters*,
SpringerBriefs in Applied Sciences and Technology,
https://doi.org/10.1007/978-981-16-6651-3_2

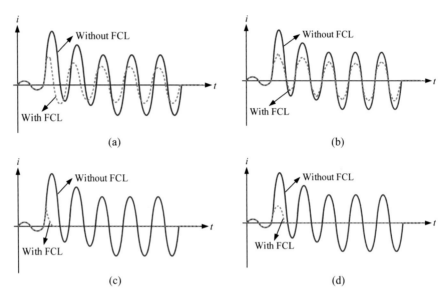

Fig. 2.1 Example fault current waveforms of different types of FCL [8]; **a** Type A1 (resistive type), **b** Type A2, **c** Type B1, **d** Type B2

The classification presented by the IEEE standard is based on current limiting behavior. It neglects the basic technologies used in the design of FCL. Most of the review papers and CIGRE and EPRI reports classify FCLs considering applied technology [1, 2, 11–12]. Accordingly, FCL structures are categorized into two groups including superconducting fault current limiters (SFCLs) and non-superconducting fault current limiters (NSFCLs). The superconducting materials are used as the main element in SFCLs. The potential projects related to FCL around the world focus on SFCL technology. Moreover, the technical feasibility of the application of SFCL in T&D systems has been demonstrated by the installation of numerous SFCLs in the real power grids for field tests or permanent applications. In NSFCL, as its name implies, the limitation duty is achieved without superconductors. NSFCLs generally have four subsets: liquid metal fault current limiters (LMFCL), positive temperature coefficient (PTC) based resistors, solid-state fault current limiters (SSFCLs), and Is-limiters. Is-limiter is a fuse-like explosive device, but it is considered as FCL technology in some reports. Figure 2.2 summarizes the different classifications of FCLs. The detailed descriptions of the potential structures of each class are presented in the next subsections.

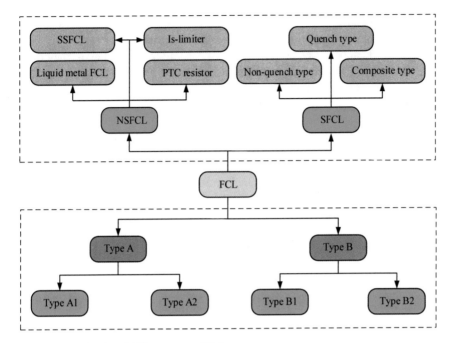

Fig. 2.2 Classification of different types of FCL

2.1 SFCL

SFCL is one of the most prominent candidates of current limitation technology. It can overcome the high short circuit level problem at both T&D systems. The superconductor is the main element in the design of SFCL. The electric resistance of superconductors becomes nearly zero in the superconductivity state. Therefore, the negligible power loss is produced during normal current conduction. The superconductivity mode appears in superconductors when three parameters of superconductor consisting of current density (J), temperature (T), and applied magnetic field intensity (H) are all lower than the critical values (J_c, T_c, and H_c, respectively). The critical values depend on the type of material. As one of the mentioned quantities becomes more than the critical value, the superconductor suddenly switches to normal mode and the electric resistance of superconductor quickly increases. The term "quench" is used in the literature when the superconductor transits from superconductivity mode to normal mode. First of all, the below explanations are also required to know about superconductors [12–13]:

- The current density of superconductors is higher than the conventional conductors such as copper and aluminum materials. It leads to a lower cross-section area at the same conditions;
- The power loss problem associated with superconductors is a lesser concern against the conventional conductors;

- The power loss of superconductors in AC applications is more than DC applications due to alternating current (called AC loss);
- Superconductors can be classified into two main groups in view of the critical temperature: low-temperature superconductor (LTS), and high-temperature superconductor (HTS). In order to achieve the superconductivity state, LTS should be significantly cooled in comparison with HTS. Since LTS has a critical temperature near absolute zero. Hence, more powerful cooling systems are required. Liquid helium is mostly used in the cooling system of LTS materials, whereas HTS materials are cooled through liquid nitrogen (LN_2) coolant. Liquid nitrogen is cheaper and safer than liquid helium. Most of the research on superconducting equipment such as SFCL, superconducting cables, and transformers are concentrated on HTS materials at now due to the above-mentioned challenges about LTS materials;
- The terms "cryogenic", "cryostats", and "cryocoolers" are commonly used in superconducting equipment. In physics, a science that focuses on the production and behavior of materials at very low temperatures is called cryogenic. Cryogenic involves temperature ranging from –273 °C (0 K) to –150 °C (123 K). Cryocooler or cryogenic cooler is an active device that is designed to cool things and reach them under cryogenic temperature. In reality, it is a refrigerator. Finally, the cryostat is a passive device that maintains things at cryogenic temperatures;
- The superconductors can be also divided into two groups consisting of Type I and Type II in terms of the magnetic property. While Type I is described by critical magnetic field intensity (H_c); two critical magnetic field intensities (H_{c1} and H_{c2}) are defined for Type II;
- Figure 2.3 illustrates the operational states of the Type II superconductor. As can be seen, the superconductivity mode is maintained when the H becomes lower than H_{c1}. In the case that H is upper H_{c2}, complete loss of superconductivity mode happens that is known as the normal state. For magnetic field intensities between H_{c1} and H_{c2}, the terms "flux-flow state" or "mixed state" are used for description. The superconductor in this state introduces higher impedance than superconductivity state but is lower than normal state. It should be noted that the HTS materials are mostly Type II.
- As illustrated in Fig. 2.4a, the weak magnetic field is expelled by a superconductor in superconductivity mode. In fact, it only penetrates into the surface of the superconductor and a magnetic shield is provided. London penetration depth describes the distance that the magnetic field penetrates into a superconductor surface. It

Fig. 2.3 Operational states of Type II superconductor

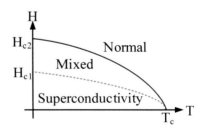

Fig. 2.4 Meissner effect; **a**
H < H$_c$, **b** H > H$_c$

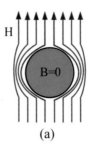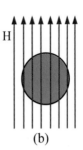

(a)   (b)

is very low in the superconductivity state. However, as shown in Fig. 2.4b, this property is eliminated when the applied magnetic field becomes more than H$_c$ (Type I) or H$_{c1}$ (Type II). This phenomenon, known as the Meissner effect, can be applied to develop current limitation technology. During fault conditions, fault current generates the strong magnetic field and the magnetic shield provided by superconductor destroys.

- Bismuth-Strontium-Calcium-Copper-Oxide (BSCCO), Yttrium-Barium-Copper-Oxide (YBCO), and Magnesium Diboride (MgB$_2$) are the HTS materials that are mostly used for developing superconducting equipment especially SFCL. BSCCO and YBCO are considered as first and second generations of HTS materials, respectively. BSSCO tube, BSSCO tap, YBCO thin films, and YBCO coated conductors are different types of HTS that are employed in SFCL structures. While both generations are maintained in liquid nitrogen, temperature ranging from 65 to 80 K. The desirable temperature for MgB$_2$ is between 20 and 30 K. In terms of comparison, interest in the application of YBCO families is more than BSCCO ones because of the development cost and utilization of silver in BSCCO. Moreover, MgB$_2$ is cheaper than YBCO and BSCCO, but its application in AC equipment is limited due to lower critical temperature and higher AC loss. The MgB$_2$ is preferred in DC application, at now.

So far, the different structures based on superconducting technology have been proposed. Figure 2.5 shows a classification for SFCL topologies [12, 14]. They have three subsets including quench type SFCL, non-quench type SFCL, and composite type SFCL. Quench type SFCL, as its name implies, limits fault current through the transition from superconductivity state to the normal state. Transition is begun by exceeding the quantity of effective parameters from critical value. All of the introduced structures in this group need recovery time after fault clearing from the HTS element. The concern associated with the recovery of the superconductor is eliminated in non-quench type SFCL. In this concept, the superconductivity state is maintained in all conditions. HTS element is mostly used in this structure instead of conventional conductors to overcome significant power loss. Finally, in the composite type SFCL, keeping the superconductivity mode depends on the type of fault. HTS elements may lose the superconductivity state during a particular fault. Therefore, the time for recovery of superconductor may need. The different structures in each group are discussed in the subsequent subsections.

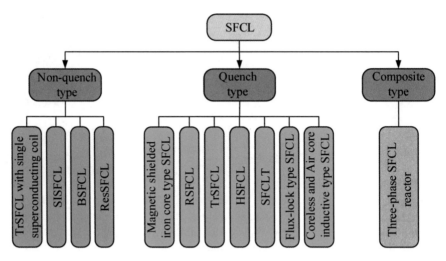

Fig. 2.5 Classification of SFCL topologies

2.1.1 Quench Type SFCL

2.1.1.1 Resistive Type SFCL

Resistive type SFCL (RSFCL) is considered as the simplest current limitation technology. It can be implemented with an HTS element. However, it may also include a shunt impedance to meet some requirements such as management of hotspot problem and excessive build-up overvoltage across RSFCL terminals, as shown in Fig. 2.6 [6]. Furthermore, the matrix-like arrangement of RSFCL (see Fig. 2.7) can be also applied for high voltage and current applications [15].

Fig. 2.6 RSFCL with shunt branch [6]

Fig. 2.7 Matrix type SFCL [12]

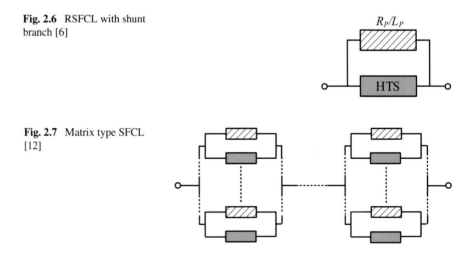

Under normal load current, the superconducting element introduces a low resistance path. In this case, most of the power loss caused by SFCL is subjected to AC loss. On the other hand, it inserts a large resistance to the circuit when a fault happens. As soon as the fault current reaches the critical value, the superconductor quickly quenches and becomes resistive. The quench time in faulty lines is not exactly the same, but it is often very short [16]. The quench leads to the temperature of the cryogenic environment starts rising due to dissipate heat by the superconductor. If this temperature maintains below the critical value during fault conditions, developed RSFCL is called as the flux-flow type [17]. Otherwise, typical RSFCL is organized. In a similar volume, the resistance of the superconductor in flux-flow mode is lower than the normal mode [18]. Therefore, in order to achieve enough limitation rate by flux-flow-based RSFCL, more superconducting materials are required. Another drawback of this type of RSFCL is the dependency of emerged resistance on instantaneous current, magnetic field, and temperature [19]. Nonetheless, the recovery time associated with flux-flow type is not considered as a challenging issue. The recovery process of another type may last for a long time. Furthermore, normal resistance of typical RSFCL appears in the shortest possible time, which can generate a dangerous voltage across RSFCL terminals. The implementation of shunt impedance (with resistive (R_P) or inductive elements (L_P)) with superconductor and transferring major part of fault current into it can modify this harmful voltage. Additionally, the shunt branch reduces the concerns related to hot spots problem during limitation function in case of placing it in the touch of superconductor (resistive element) or coaxial arrangement (inductive element) [20].

The focus of most researchers and companies on this type of FCL has provided significant theoretical and experimental information about it. Besides, it has a simple working principle, compact design, and fast response time. So far, it was constructed in different capacities and installed in the real power system to assess its performance in the long-term or even use permanently [6, 12, 21].

2.1.1.2 Hybrid Type SFCL

The combination of RSFCL with the current limiting part (CLP), a mechanical switch, and or the semiconducting switches can be also used to overcome the fault current. This principle, which is known as hybrid type SFCL (HSFCL), has attracted attention in the recent decade. Fault current is mainly limited by CLP, instead of superconductor in this concept. The superconductor only limits fault current in the first moments and then commutates it into CLP. Such operation is favor due to the following reasons:

- The minimization of required HTS materials is achieved due to reduction of HTS limitation time;
- The lower amount of HTS material results in the reduction of the AC loss and initial costs;
- The recovery process of HTS element reduces; meaning that the capacity of the cooling system can be minimized;

Fig. 2.8 CB-based HSFCL [24]

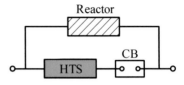

- Maintenance costs are also minimized;
- CLP can withstand fault current for a longer time;
- Due to this fact that the voltage stresses are subjected to mechanical switch, not the HTS element, more convenient conditions are provided for high voltage applications of HSFCL.

More than 10 innovative structures for these concepts have been proposed in the literature. Some of them are restricted to depth computer analysis, but some have been significantly developed and even installed in the power system. The topology shown in Fig. 2.8 can be considered as the mature type of HSFCL. The successful operation of this type has been demonstrated by several prototypes and live-grid installation [22–24]. HTS element, CB, and reactor (as the CLP) fulfill the configuration of this HSFCL. The resistor can be also used instead of a reactor in CLP if the dissipated heat by the resistor is well controlled. Normal load current mainly passes through the unrestricted path provided by HTS and CB. They have a negligible impedance in comparison with the shunt reactor. However, when the power system experiences a fault, superconductor quenches and exhibits the large electric resistance. The result of this behavior is to commutate the major part of the current into the shunt branch. The limitation function of this SFCL during fault conditions can be generally divided into two stages; prior and after CB operation. The equivalent impedance of the reactor and electric resistance of the HTS element limits fault current before CB operation. Subsequently, the limitation duty is completely done by the reactor as soon as the arc between contacts of CB is extinguished during the interruption. More limitation is also achieved in this step. Equation (2.1) mathematically expresses the above statement [25]. t_{quench} and t_{CB} in this equation are defined as the starting time of superconductor transition and the action time of CB, respectively.

$$Z_{HSFCL} = \begin{cases} 0 & t < t_{quench} \\ R_{HTS}(t) \, || Z_{Reactor} & t_{quench} \leq t < t_{CB} \\ Z_{Reactor} & t \geq t_{CB} \end{cases} \qquad (2.1)$$

HSFCL can be also developed by the implementation of a fast switch instead of CB. Figure 2.9 represents one of the suggested fast switch-based HSFCL [26]. HSFCL includes three paths and four components comprising of HTS element, fuse, CLP, and fast switch. Driving coil, main contact, and auxiliary contact also complete the parts of the fast switch. The working principle of this HSFCL can be described in four steps. In the first one, normal state current would flow through the series arrangement of HTS and main contact. Once short circuit current leads to quenching in the

Fig. 2.9 Fast switch-based HSFCL [26]

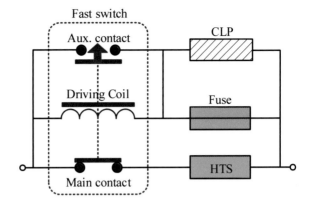

superconductor, the resistance of superconductor transfers the main part of current into shunt branch consisting of driving coil and fuse. Then the strong magnetic field provided by driving coils leads to produce the adequate electromagnetic repulsion force to energize fast switch. So, the states of main and auxiliary contacts change. Main and auxiliary contacts experience the open and closed conditions, respectively. These events transfer the entire fault current into driving coil and auxiliary contacts, and the HTS element enters in the recovery mode. Finally, the fuse melts, and CLP limits fault current. All the above steps happen just a few milliseconds. Therefore, the fault is moderated before the first peak of fault current.

While this type of HSFCL meets some requirements such as low recovery time, automatic operation, low development, and maintenance costs, the fuse should be replaced after each operation. Furthermore, the limitation duty of HSFCL depends on the melting time fuse. In order to alleviate the problems associated with the above structure, several evolved models have been proposed. Figure 2.10 illustrates some of the suggested structures [27–29]. The elimination of fuse in these models is the main difference that leads to quickly return HSFCL in service after each operation. However, the additional functions are also provided by some of them. For example, the concept shown in Fig. 2.10b does not limit fault current in the first half cycle to provide flexibility in the operation of the protective relays. The ability to control the fault current reduction ratio or even interrupt it can be achieved by the structure shown in Fig. 2.10d.

2.1.1.3 Magnetic Shielded Iron-Core Type SFCL

Magnetic shielded iron-core type SFCL uses the Meissner effect to limit fault current [30]. Figure 2.11 illustrates its configuration. It consists of an iron-core, an HTS cylinder, and a conventional winding. HTS element is placed between the iron-core and conventional winding [31]. It does not have the physical terminal and is completely isolated from the power system. Therefore, the concerns related to the

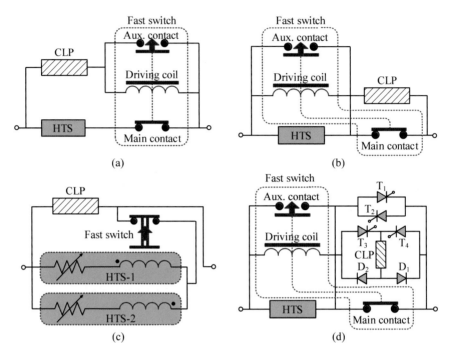

Fig. 2.10 Some of the evolved models for fast switch-based HSFCL; **a** HSFCL with first half cycle limiting operation [27], **b** HSFCL with the first half cycle non-limiting operation [27], **c** HSFCL with two HTS elements [28], **d** HSFCL with ability to control fault current [29]

Fig. 2.11 Configuration of magnetic shielded iron-core-type SFCL [6]

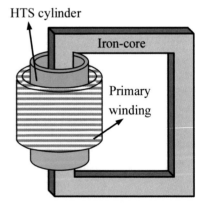

physical integration of superconductor with the power system and subsequently additional required components such as current leads are disregarded. The conventional winding, which is implemented with the copper conductor, is connected in series with the power system. In terms of operating principle, magnetic shielded iron-core type SFCL is similar to a series transformer that the primary and secondary windings are constructed by copper conductor and superconductor, respectively. The turn number of the secondary side is unit and this side works like short circuit winding.

During normal load current, the Meissner effect causes acting the HTS cylinder as a shield against the produced magnetic field by copper winding. Hence, the magnetic flux of iron-core is almost zero, and the provided impedance on the primary side is low. The total voltage drop of SFCL is subjected to copper winding and leakage reactance in this mode. However, the fault current causes the copper winding to generate enough magnetic field to eliminate the Meissner effect and destroy the superconductivity state. Thereafter, the elimination of magnetic shield and penetration of the magnetic field into the iron-core is expected. Finally, the fault is well controlled by the resistance of superconductor and introduced reactance.

Although this type of SFCL is better than RSFCL in some technical aspects such as hotspot problem and no need for current leads, utilization of iron-core makes SFCL heaviness and large. Furthermore, in order to prevent magnetic coupling and disturbance, the non-conductive materials should be used in the design of cryostat and insulations of SFCL. At present, the activities in the development of this structure are very low, but the modified versions such as coreless type SFCL (it will be explained in the next subsection) are still ongoing.

2.1.1.4 Coreless and Air-Core Inductive Type SFCL

A coreless model and air-core inductive type SFCL have been presented to eliminate the heavyweight and large size of magnetic shielded iron-core type SFCL [32–33]. The structure and operating principle of this concept are completely the same as magnetic shielded iron-core type SFCL. The primary winding in this type works like an air-core reactor. In the same ratings, the leakage reactance of the coreless/air-core types is lower than the iron-core type because of the minor magnetic permeability of air. So, the voltage drop of SFCL reduces during normal conditions. Regardless of this positive aspect, the mutual coupling between windings is low when no iron-core is used. In order to achieve the sufficient mutual coupling between windings and reasonable current limiting percentage during fault conditions, three techniques can be implemented in the design phase; reducing the distance between HTS cylinder and copper winding, increasing the number of turns of primary windings, and finally using more superconducting material to obtain higher electric resistance. The use of superconducting material individually is not cost-effective and can be combined with two remaining techniques. In the coreless type, the distance between the primary winding and HTS elements is minimized by placing them at the cryogenic temperature. It is evident that the power loss in this type is considerable, and the high capacity cooling system is required. Furthermore, current leads should be also implemented. In air-core type SFCL, the primary side operates in ambient temperature and the adequate mutual coupling is provided through the proper choice of turns of conventional winding. Both types of mentioned SFCLs are in development stages. Moreover, several prototypes in medium capacities have been constructed [34–35].

Fig. 2.12 SFCLT structure
[36]

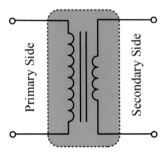

2.1.1.5 Superconducting Fault Current Limiting Transformer

Superconducting fault current limiting transformer (SFCLT) is a superconducting transformer with fault current limitation capability. The structure of SFCLT is shown in Fig. 2.12 [36]. The superconducting materials are used to provide primary and secondary windings of the transformer. Both windings are in liquid nitrogen. Under normal conditions, SFCLT operates like a step-up or step-down conventional transformer, but with lower voltage drop, higher efficiency, and lower power loss. When a fault takes place, the fault current causes the superconducting windings to become resistive and subsequently fault current is suppressed by the provided resistance and leakage reactance. A device with two functions is an interesting idea. The combination of units decreases the total weight, required space and development, and maintenance costs.

Although in some papers SFCLT has been designed with the capability of recovery under load characteristic [37], the recovery process is still a challenging issue. Unwanted quench causes the transformer function of SFCLT, which does not work properly. Conventional transformers should be able to tolerate the fault current for 2 s regarding the IEC standard [38]. Although this time is also expected to be considered in the design of SFCLT, superconductor may damage due to the dissipated heat. Several prototypes based on these concepts have been constructed in different capacities, but no trial installation has been reported [39–40].

2.1.1.6 Transformer Type SFCL

Figure 2.13 depicts the schematic of transformer type SFCL (TrSFCL). It includes a conventional transformer and an HTS element in each phase [41]. While the primary side of the transformer is connected in series with the circuit, the HTS element is integrated into the secondary side. The current in the secondary side is lower than the primary one. Prior to fault occurrence, resistance and leakage reactance of the transformer only make the voltage drop across SFCL terminals. Since the impedance of the primary side of the transformer is nearly zero due to the short path provided by the superconductor. As a fault changes the current of the secondary side of the

Fig. 2.13 Schematic of
TrSFCL [41]

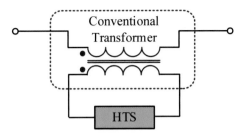

transformer, the electric resistance of superconductor returns. It leads to a large impedance seen from the primary side that limits fault current.

The current of the superconducting element in the TrSFCL is lower than the system current. It conveniences the management of the hotspot problem, the build-up voltage across the HTS element, dissipated heat by superconductor during the quench, and recovery process. Furthermore, impulse waveforms such as lightning have a minor effect on superconductor due to the coupling of the superconductor with the secondary side of the transformer. However, the use of the conventional transformer causes a significant power loss. The waveform distortions and low current limiting percentage may happen when the iron-core is saturated by fault current [42]. Most of the research in this type and its modified versions have focused on computer analysis and low-capacity experimental tests [43–45].

2.1.1.7 Flux-Lock Type SFCL

Figure 2.14a displays the basic structure of the flux-lock type SFCL family [46]. It includes three conventional windings wound on an iron-core, superconducting element, magnetic field coil, resistor, and capacitor. The series combination of coil 2 and the superconducting element is integrated with coil 1 in parallel. The terminals of coil 3 are also connected to the series arrangement of the resistor, capacitor, and

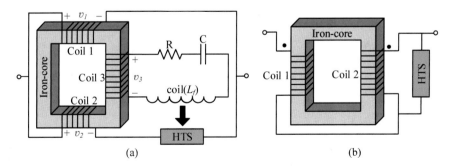

Fig. 2.14 Flux-lock-type SFCL concept; **a** basic structure [46] **b** flux-lock type with the series connection of coil 1 and coil 2 [50]

magnetic field coil. The voltage across each coil can be determined by the following equations [46]:

$$v_1 = n_1 \frac{d\phi}{dt}, \ v_2 = n_2 \frac{d\phi}{dt}, \ v_3 = n_3 \frac{d\phi}{dt} \qquad (2.2)$$

where ϕ is the linkage flux of iron-core. n_1, n_2, and n_3 are also the number of turns of coils 1 to 3, respectively. The turn's number of each coil is different from others. Regarding the winding direction of coils 1 and coil 2 (additive polarity and subtractive polarity), the considered positive or negative signs for v_2 is determined. When the winding direction of coil 1 is the opposite of coil 2, the negative sign is used. It should be noted that the winding direction of coils 1 and 2 can affect the current limiting percentage of SFCL [47].

In normal case, concerning the electric resistance of the HTS element is nearly zero, v_1 is equal to v_2 ($v_1 = v_2$). Hence, the following equation can be extracted in this case.

$$v_1 = v_2 \Rightarrow v_1 - v_2 = 0 \qquad (2.3)$$

By using Eq. (2.3), gives:

$$(n_1 \pm n_2) \frac{d\phi}{dt} = 0 \qquad (2.4)$$

The turns' number of coil 1 and coil 2 is not similar. In this regard, $d\phi/dt$ should be zero to satisfy the above equation. This means that the iron-core experience the constant linkage flux during normal conditions and the induced voltage in all coils is zero. Hence, the voltage across SFCL terminals is only subjected to the resistance of conventional windings. As soon as the fault current causes appearing the electric resistance of superconductor, Eq. 2.4 is no longer valid and linkage flux would alter as the function of time. Inducing the voltage in coils and suppressing fault current are the subsequent results of this event. Furthermore, the appeared voltage across terminals of coil 3 causes flowing current from the magnetic field coil. The produced magnetic field by these coils helps the superconductor to emerge a larger resistance and thereby achieve more current limitation. The stability of the superconductor during fault conditions is improved by using a capacitor. However, it may have a negative impact on the current limiting behavior of SFCL. Because it increases the flowing current from coil 3 and consequently reduces the linkage flux of iron-core. This capacitor and magnetic field coil are so designed that the series resonant circuit is formed. The use of a resistor in the resonant circuit keeps the resonant circuit against the large current.

There are also several modified structures based on this type of SFCL. In [48–49], the authors have been suggested a control circuit consisting of semiconductor switches and tap changer instead of the utilization of resistor and capacitor. The control of fault current is also viable by implementing these strategies. The elimination of coil 3 and the use of the series connection of coil 1 and coil 2 (without considering the HTS element) are also investigated in [50] whose schematic is shown in Fig. 2.14b. Coil 1 and coil 2 have the same turns in this concept. The detail of most of the derivative structures of flux-lock-type SFCL can be found in [51].

2.1.2 Non-Quench Type SFCL

2.1.2.1 Saturated Iron-Core Type SFCL

Saturated iron-core type SFCL (SISFCL) can be considered as one of the potential types of SFCL to control the fault current. The technical feasibility of the application of this type has been demonstrated through the depth computer analyses, experimental studies, and field tests in different ratings. Figure 2.15a illustrates the basic structure of conventional SISFCL. HTS coil, iron-cores, DC power supply, and conventional coils (also called AC coils or copper coils) are the main components of the SISFCL. Two series of conventional coils with the same turns and opposite direction are implemented in each phase of SISFCL. However, the number of HTS coils depends on the design in three-phase applications. While three DC coils are required when three single-phase units are applied (bank form), a common HTS coil is used in the unified design. The role of the DC coil is to generate a strong magnetic field to drive the iron-cores into deep saturation in normal state. It is energized by the powerful DC source. The use of MgB_2 instead of other more expensive HTS materials is feasible in this structure. It is due to the fact that the superconductor works in DC mode and the AC loss is not a concern.

The non-linear magnetic characteristic of ferromagnetic material (B-H curve) plays the main role in the limitation function of SISFCL [52]. Under normal operation of the power system, iron-cores work in the saturation region. The magnetic fields generated by AC coils in this state can change the operational point of iron-cores in

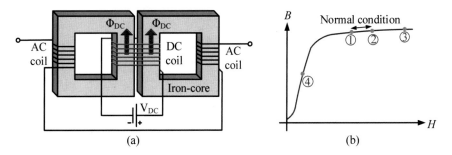

(a)   (b)

Fig. 2.15 SISFCL concept; **a** Single-phase structure [52], **b** B-H curve of iron-core and its operational points under different conditions of SISFCL

a few ranges. The interval between points labeled as "1" and "2" in Fig. 2.15b shows the variation of flux density of each iron-core in negative and positive half cycles of load current. As known, the magnetic permeability of ferromagnetic material in the saturation region is low. Hence, the provided inductance by SISFCL is not high enough to impact load current. The total voltage drop of SISFCL is related to the inherent resistance of conventional windings. As a fault significantly changes the amplitude of current, the produced magnetic field by AC coils can alternately push out operational points of both iron-cores to linear region. The points labeled as "3" and "4" in Fig. 2.15b specify the status of iron-cores during fault conditions. Indeed, the use of two iron-cores guarantees the operation of SISFCL in both positive and negative half-cycles. Anyway, regarding the magnetic permeability of iron-core in this region is more than that of the saturation state, SISFCL enters a large inductance in the circuit and fault current is limited without any delay.

SISFCL offers some benefits such as quick response, low recovery time, and feasibility to apply in different voltage ratings. Several successful practical projects have demonstrated the efficiency of this SFCL. The live-grid tests of 35 kV/1.5 kA, 15 kV/1.25 kA, 220 kV/0.8 kA SISFCL have been reported in refs [52–55]. Furthermore, the significant step to design 500 kV/300 MVA has been done [56]. However, conventional SISFCL has a bulky design and heavyweight in comparison with other topologies. For example, the weight of the 220 kV/0.8 kA SISFCL project was about 120 ton. The weight and dimension of equipment are very important factors especially in indoor applications due to space limitation. The non-linear behavior of the B-H curve of the iron-core may cause voltage distortion during fault conditions. In order to produce the DC magnetic field, the turn number of the coil and required DC current should be high. Therefore, a high-capacity DC power supply is needed. The reliability of this power supply is important [57]. Its unavailability converts SISFCL into the series reactor with high voltage drop. The build-up voltage across DC coil during the fault event may cause the failure of dielectric insulation. To avoid the above problem, the utilization of a control circuit consisting of high voltage insulated gate bipolar transistor (HVIGBT) and piezoresistor has been suggested to remove the DC coil [58]. Despite the complexity, the above tact also improves the current limitation capability of SISFCL due to reactor-like operation. Finally, the provided inductance and subsequently the fault current reduction ratio of SISFCL change by the magnitude of fault current [59]. In extreme fault current, both iron-cores (in each phase) may fall in saturation region via the produced magnetic field by AC coil. The negligible current limitation is the consequence of this case.

It should be noted that the SISFCL concept does not restrict the above-mentioned structure, and there are some other derivatives. In fact, the discussed SISFCL was the basic structure of the SISFCL family. The utilization of conventional conductor or permanent magnet and or their combination instead of superconductor have been also investigated in some topologies [60–61]. Additionally, open core and open closed structures (shown in Fig. 2.16) are the innovative concepts that have been proposed and tested [62–63]. All of SISFCL derivatives work like the conventional structure. They are only trying to conquer some of the drawbacks of conventional topologies.

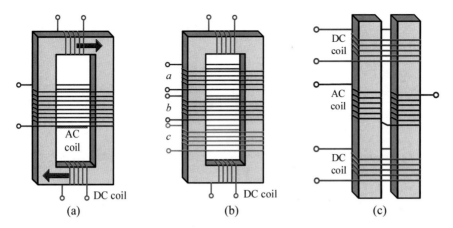

Fig. 2.16 Open closed and open core SISFCL concepts; **a** Structure of single phase open closed SISFCL [62], **b** Three-phase open closed SISFCL structure, **c** Open core SISFCL [63]

2.1.2.2 Resonant Type SFCL

Resonant type SFCL (ResSFCL) manages fault current based on the resonance phenomenon. In an electrical circuit consisting of a parallel or series connection of a capacitor and an inductor, the electrical resonance happens at a certain frequency, which is known as the resonant frequency. In this frequency, the reactance of the inductor and the capacitor is the same in terms of magnitude. Figure 2.17a shows the structure of ResSFCL. A superconducting coil and capacitor are used in this type [64]. Line frequency (50 Hz or 60 Hz) is considered as the resonant frequency. Prior to fault occurrence, ResSFCL provides an unrestricted path to flow load current. However, it enters in limiting mode as a fault takes place. The smooth increase in fault current (Low DC component) and no waveform distortions are expected from ResSFCL operation. It is due to the dynamic behavior of the resonant circuit that can be theoretically concluded from the below equation [64].

$$i(t) = V_m[\frac{1}{Z_L}\sin(\omega t + \alpha - \phi) + \frac{1}{2\omega L_{SFCL}}\sin\alpha\sin\omega t + \frac{1}{2L_{SFCL}}t\sin(\omega t + \alpha)]$$

$$(2.5)$$

where α and $Z_L\angle\phi$ are voltage angle at fault inception time and aggregate impedance of transmission line and load, respectively. The use of ResSFCL introduces a fault current with three terms as observed in (2.5). The oscillation of fault current at line frequency is concluded from the first and second terms of this equation. Although the last term is also an oscillation one, the rate of rise of fault current in this term is low regarding the constant slope ($V_m/2L_{SFCL}$). The limitation rate depends on the slope and subsequently L_{SFCL}. The lower slope would be provided by higher inductance.

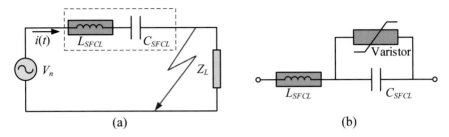

Fig. 2.17 Resonant-type SFCL [64]; **a** The basic configuration of the series LC circuit, **b** Resonant-type SFCL with varistor

Simple structure, no-fault detection system, quick response, no waveform distortions, and gradual increase of fault current are the appropriate advantages of ResSFCL. No cryogenic system is also achievable in case of using the conventional conductor to develop coil. Nevertheless, ResSFCL is not capable to limit fault current for a long time. It eventually reaches a maximum value, which can put at risk the power system. The capacitor size is expected to be large in the HV system. Unwanted quench may occur in excessive fault current. ResSFCL has a significant adverse effect on interrupting the characteristics of CB [65–66]. The build-up voltages across the capacitor and coil may be significant. One way to reduce the concerns related to this overvoltage is to add a varistor (metal oxide varistor) with a capacitor in parallel, as depicted in Fig. 2.17b [64]. In addition to the protection function, the limitation of fault current in steady state is obtained after clipping the voltage. Referring to the suggested range in refs [64, 66], the protective level of MOV should be considered between 1.8 and 2.5 per unit of the peak of provided voltage across the capacitor terminals under the nominal line current flow. The varistor should be able to absorb a large amount of energy during fault conditions. The above-mentioned drawbacks reduce the interest in the development of this SFCL.

2.1.2.3 Bridge Type SFCL

Bridge type SFCL as illustrated in Fig. 2.18a, b is divided into two main structures: single phase and three phase [67]. Power diodes and superconducting coil (DC superconducting reactor) are the common components in both topologies. The connection of diodes rectifiers AC input current. Three-phase structure needs for three conventional transformers. The transformer may be also used in the single-phase structure for high voltage applications. In contrast to the current rating, it can reduce the voltage rating of diodes. DC power supply, which is applied to charge inductor before energizing SFCL, is optional. The amplitude of I_{DC} should be greater than the normal peak of AC input current to achieve the appropriate operation. It is neglected in some studies by suggesting other innovative methods [3].

The operating principle of the topology shown in Fig. 2.18a is as follows. In normal cases, the current of DC reactor causes all power diodes to be in conducting

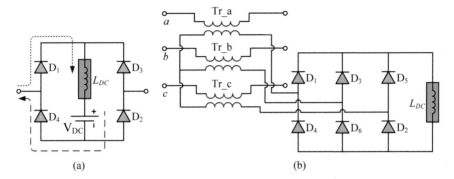

Fig. 2.18 Bridge-type SFCL [3]; **a** Single-phase configuration, **b** Three-phase configuration

state. Moreover, as known, the voltage across an inductor is zero in DC current after charging. Therefore, the voltage drop of SFCL is subjected to the on-state voltage of diodes. It can be also compensated using the DC power supply with an amplitude equal to the provided voltage drop. The change of AC peak current due to fault occurrence is expected to be higher than I_{DC}, and the limitation task is ongoing. The power diodes regularly are turned off regarding the positive and negative half cycles of fault current. In the former half cycle, the diodes labeled as D_3 and D_4 in Fig. 2.18a automatically become off, and fault current is established through the path entailing of D_1, DC reactor, DC power supply (if any), and D_2. Fault current also flows through D_3, DC reactor, DC power supply, and D_4 for later half cycle. DC reactor is spontaneously inserted in the power system in both cycles and limits fault current. The limitation duty would last until fully charging of DC reactor or operation of other protective devices. Indeed, when inductor charges to new DC current, no voltage drop is provided and fault current returns to unlimited value. Longer charging time of inductor and consequently, limitation time are achieved by considering the higher value for this element.

This kind of SFCL is able to reduce the DC component of fault current without the concerns about surge voltage and waveform distortions. Indeed, fault current smoothly rises. The electric resistance of the superconductor is low for all circumstances. Various types of superconductors can be used in the construction of the DC reactor, especially MgB_2. The mature power electronics device offers the possibility to stack or parallel them for high voltage and current applications. However, this SFCL like other types has drawbacks. The conducting loss and voltage drop of power diodes should be managed. In practice, it is hard to use the DC power supply in this topology. Although the current of superconducting element theoretically is pure DC, the ripple at twice the line frequency is created in real design. This ripple generates the voltage drop and AC loss. It decreases when the DC reactor with higher inductance is used, but the technical and economic aspects should be elucidated. Finally, additional protective devices should be also considered. Since the power diode would be damaged due to great dissipated heat.

While the single-phase structure needs for three DC superconducting reactors in the three-phase application, a superconducting coil is implemented in three-phase topology (also called transformer-isolated diode bridge type SFCL) as observed in Fig. 2.18b. In this type, a full-wave rectifier causes the DC output current to flow through the inductor. Similar to the previous structure, this type has three operational modes consisting of start-up, normal state, and fault conditions. In the first stage, the DC reactor is smoothly magnetized through the power system. As the inductor was fully charged, the variation of inductor current becomes zero and it operates like a short-circuited path. Therefore, the total voltage drop in the primary side of transformers is only related to the winding resistance and leakage reactance. Finally, when a fault happens, SFCL quickly activates. However, SFCL exhibits two behaviors against this incident. In the situation that the symmetrical fault appears in the power system, the fault current starts to rise smoothly without any surge waveform. The fault current eventually reaches to unlimited level after charging superconducting coil to new DC output current. Indeed, SFCL only buys the additional time for operation of other protective devices during the occurrence of the symmetrical fault. As the asymmetrical faults appear in the power system, the steady-state fault current is also suppressed. Because referring to the amplitude of fault current is not the same in three phase, the current passing through the DC reactor is not constant. As a result, the voltage appears between inductor terminals and following SFCL. The effect on the operation of a healthy line under asymmetrical fault is the important drawback of this topology. Moreover, the utilization of three conventional transformers is not an interesting idea. Many research studies have been carried out in the field of both structures and their derivatives [3].

2.1.2.4 TrSFCL with Single Superconducting Coil

TrSFCL with a single superconducting coil contains three single-phase conventional transformers and an HTS coil as shown in Fig. 2.19 [68]. The secondary sides of transformers are connected to a superconducting element. The use of one HTS for a three-phase application reduces the amount of required superconductor and cooling system capacity. Under safe conditions, the current of the HTS element is minor due to the symmetrical operation of the power system. Hence, AC loss is nearly zero. Additionally, no voltage drop is provided on the secondary side of the transformers. It is similar to the case in which the secondary sides of transformers are shorted. As a result, the primary sides of transformers experience the same conditions. It implies that the SFCL has a low adverse impact on the normal operation of the power system. Winding resistance and leakage reactance of transformers only produce the voltage drop across SFCL terminals in this case. As soon as the power system experiences the asymmetrical fault event, the current of the HTS element is no longer negligible. Asymmetry leads to introduce the voltage drop in the secondary sides of transformers and subsequently primary sides. From this time, the current would be limited. According to the operation principle, although the superconductor

Fig. 2.19 Transformer-type SFCL with single superconducting coil [68]

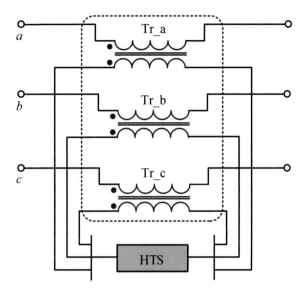

remains in superconductivity state in all conditions; however, SFCL cannot overcome the symmetrical faults, which is a noticeable drawback. SFCL only exhibits a considerable limitation during single-phase to ground fault. Moreover, the weight and size of SFCL seem to be high due to the application of three conventional transformers. The saturation of iron-core may also affect the limitation capability of SFCL and current and voltage waveforms. There are no new activities for the development of this kind of SFCL.

2.1.3 Composite Type SFCL

2.1.3.1 Three-Phase SFCL Reactor

Three-phase SFCL reactor uses three superconducting windings to fulfill the limitation duty. The turns of three windings are the same and are wound on an iron-core, as illustrated in Fig. 2.20a [69]. They are connected in series with the power system. The direction of windings is so that the total magnetic flux in the iron-core is minor in normal state. Hence, the provided impedance by SFCL and its voltage drop is expected to be minor. In the event of a fault, the superconducting windings may lose its superconductivity state. The superconductivity maintains in a single phase to ground fault. As soon as the current of one phase of the system starts rising, the total linkage flux is no longer negligible. Therefore, SFCL introduces a large reactance in the power system and subsequently limits the fault current. The recovery time associated with a superconductor is not a challenging issue for a single-phase to ground fault. However, this problem exists when the power system encounters with

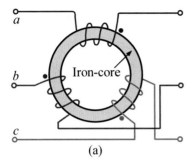
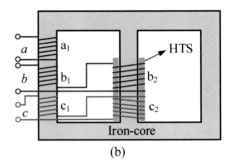

(a)   (b)

Fig. 2.20 Composite-type SFCL; **a** Three-phase SFCL reactor [69], **b** The proposed structure in [70]

other types of fault. Since the transition from superconductivity state to normal state happens in these circumstances to limit fault current. Nevertheless, the initial cost of this SFCL would be high due to the requirement of a large amount of superconductor. Besides, the healthy lines may be impacted during asymmetrical faults by SFCL operation.

Some above-mentioned problems related to the previous structure can be mitigated through the suggested topology in ref.[70]. As illustrated in Fig. 2.20b, an iron-core, five conventional windings, and a superconducting cylinder are the main components of this type. Three windings labeled as a_1, b_1, c_1 are placed at a leg (left leg in Fig. 2.20b) and windings b_2 and c_2 are wound around in the center leg. Windings are similar in terms of turns and are integrated in series with the circuit. Only winding c_2 has the opposite direction in comparison with others. The superconducting cylinder acts as the magnetic shield and prevents the magnetic field produced by windings b_2 and c_2 from penetrating into iron-core. Under normal load current, the summation of generated magnetic fields by windings a_1, b_1, and c_1 is almost nothing in left leg due to balanced conditions. Furthermore, the superconducting cylinder does not allow the magnetic field provided by winding b_2 and c_2 reaching to the iron-core. As a result, the negligible magnetic field is available in the iron-core. It implies that the total impedance of SFCL in normal conditions is due to the inherent resistance of winding that is not significant.

Two working principles for this SFCL can be expressed during fault conditions. The magnetic field in the left leg of the iron-core would be strong when an asymmetrical fault happens. It leads to insert the large impedance in the circuit and consequently current limitation. However, the superconducting cylinder loses its superconductivity state under symmetrical fault and in turn magnetic shield vanishes. So, the harmful fault current is restricted through the introduced reactance by windings. Regarding the mentioned working principle, the superconductor only needs the recovery process during this type of fault. As known, the probability of the occurrence of the symmetrical fault is lower than other types. SFCL also uses one superconducting element instead of three in this structure. The lower initial and maintenance costs of superconductors are, therefore, achieved. Finally, the concerns related to the impact of this SFCL on a healthy line under asymmetrical faults are lesser than the previous structure. Yet, the focus on the two aforementioned structures has not been an important objective within researchers.

2.2 NSFCL

2.2.1 LMFCL

As its name implies, the liquid metals are used to perform the limitation duty in this concept. Liquid metals have a low melting point and maintain the liquid at room temperature. The electrical conductivity of liquid metals is more than most of the solid metals. Therefore, power loss caused by LMs is lower, which is worthy properties. Mercury and alloy of Gallium, Indium, and Tin (Called Galinstan or GnInSn) are two well-known liquid metals. Mercury is a toxic liquid and attracts less attention in LMFCL development; whereas, the utilization of alloy of Gallium, Indium, and Tin in this application is more due to non-toxic and lower stream pressure properties. The melting temperature of this alloy is around 10 °C. The working principle of most LMFCL structures is based on the "self-pinch effect" mechanism. When a considerable current flows through the ferromagnetic liquid (here, the liquid metal), a self-generated magnetic field causes the constriction of liquid metal. Indeed, it compresses and its effective cross-section area decreases. The further current causes more constriction and finally evaporation of liquid metal and producing the arc. This arc is used to provide the voltage drop and limit fault current [71]. The generic topology of LMFCL is shown in Fig. 2.21 [72].

The insulating walls (or Spacer) divide the inner space of the enclosure into several subsections. Then, the inner space is partly filled with liquid metal. This arrangement generates many series short arcs, meaning that enough overall voltage drop across LMFCL is provided and thereby fault limitation task is realized. The constriction of liquid metal appears in implemented narrow channels on the insulating wall in which current passes through them. Strong non-homogeneous current density distribution is

Fig. 2.21 The conceptual structure of liquid metal FCL [72]

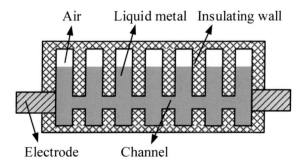

Fig. 2.22 Geometrical unstable magnetic field [72]

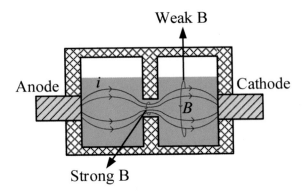

the consequence of this structure [72], as observed in Fig. 2.22. It causes the magnetic field instability, and igniting the self-pinch in the narrow channel and eventually arc production. Arc ignition, arc expansion, and arc elongation along the channel (touching anode and cathode electrodes) make three stages of arc evolution [73].

Fast response, automatic operation, automatic recovery, and simple structure are the advantages offered by LMFCL. However, until now, there is not any appropriate theory for the critical current of liquid metal. Furthermore, the efficiency of this FCL strongly depends on the pre-arcing time and peak value of voltage [71].

There is another topology for LMFCL, which limits fault current without pinch effect phenomenon [74]. The schematic design of the proposed FCL is shown in Fig. 2.23. The structure is formed from two electrodes, liquid metal and graphite resistance. Two electrodes are connected to the power system, liquid metal acts as a bridge between two electrodes. Graphite resistance also plays the current limiting role. In normal conditions, the current flows through two electrodes and liquid metal. In the presence of an external magnetic field (B), the Lorentz force (F_{mag}) exerts to liquid metal. However, the load current is not able to move liquid metal. Since the Lorentz force is less than gravity force. In the case of the power system experiences fault, the fault current considerably increases Lorentz force and overcomes the gravity force. It leads to move liquid metal. As soon as the liquid metal is separated from

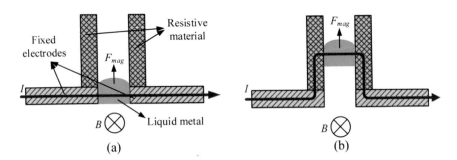

Fig. 2.23 Schematic of proposed LMFCL in [74]; **a** Normal conditions, **b** Fault conditions

the electrodes, the current is commutated into graphite resistor and fault current is limited. The stored energy in graphite resistance during operation depends on voltage and current levels. Depending on the liquid metal dynamic, the reaction time of this structure is between 100 μs and 1 ms. The detailed analysis related to this LMFCL can be found in [74].

2.2.2 PTC Resistor

Positive temperature coefficient (PTC) resistors can be also used to protect the power system against fault current. The electrical resistivity of PTC materials depends on temperature. For the temperature above the critical value (T_C), PTC material introduces high electrical resistivity, as shown in Fig. 2.24. The PTC-based FCLs use this characteristic for performing limitation duty. In normal load current, the cold PTC provides an unrestricted path. When the current starts to rise due to fault occurrence, PTC is heated and inserts high resistance to overcome the fault current. Ceramic, composite polymer, and pure nickel are several materials, which can be used to construct PTC resistor [75–77]. Nickel is a metal with a high-temperature coefficient, but the large mass of Nickel is required to avoid arriving temperature to the melting point. At now, many research studies focus on composite polymer and ceramic due to the high-temperature coefficient, high thermal capacity, and rapid changes in resistance with temperature [76]. Resettable PTC, which is also called polymeric PTC (PPTC), is widely used for protection of circuit against harmful currents. However, both types are commercially available for electronic applications (low voltage and current applications). The application of PTC resistors in the power system is restricted by nominal voltage, nominal current, breakdown voltage, and cooling characteristics. Furthermore, in the power system application, PTC should be able to absorb the energy of high short circuit current. This problem increases the cost and size of the PTC resistor. The series, or parallel, or series–parallel (matrix type) connections of PTC resistors (shown in Figs. 2.25, 2.26 and 2.27) are the several suggested ideas to overcome these restrictions [75, 77]. However, these ideas also encounter some problems. The characteristics of PTC resistors are not similar, causing the unstable distribution of current and voltage in parallel and series structures, respectively. In extreme case, this can damage PTC resistors. The

Fig. 2.24 Electrical conductivity of PTC as the function of temperature

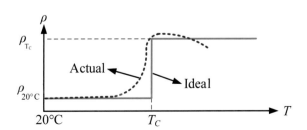

Fig. 2.25 Parallel
connection PTC resistor for
high current applications

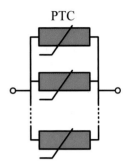

Fig. 2.26 Series connection
of PTC resistors for high
voltage applications [75]

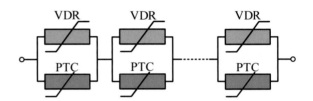

implementation of shunt voltage-dependent resistor (VDR) with PTC resistor can suppress build-up overvoltage across the PTC elements in series connection. The series–parallel connection of the PTC resistor also reduces the unstable distribution of current.

Furthermore, a structure based on the combination of PTC resistors with mechanical and semiconducting switches has been also suggested in [78]. As depicted in Fig. 2.28, it has three distinct paths, which are connected in parallel. In normal conditions, all switches are in conducting state, and load current carries through the path A due to lower resistance. During fault occurrence, the ultra-fast mechanical switch opens within a few hundred microseconds and transfers current to parallel path consisting of GTO bridge and fast switch. The diode bridge provides one direction current for GTO. When fault current in this path becomes zero, the fast switch opens without arcing and in turn, the fault current is commutated to PTC and is limited by the provided resistance by PTC. Finally, the load switch interrupts the fault current.

Fig. 2.27 Matrix structure
of the PTC resistor for high
voltage and current
applications [77]

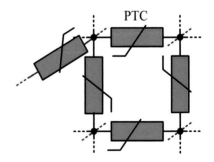

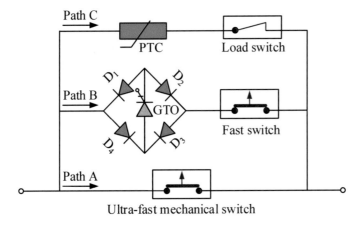

Fig. 2.28 Combination of PTC resistor with mechanical and semiconducting switches [78]

In this structure, path B only helps to commutate current from path A to path C in favor conditions. The utilization of three mechanical switches and the need for fault detection system are the disadvantages of this FCL.

2.2.3 Is-Limiter

Is-limiter is a type of a commutation-based fuse limiter that is developed to solve the problem associated with the rated current restriction of high voltage fuses [79]. Thermodynamic considerations do not allow to design an HV fuse with the high rated current because of the load current directly flows through the melting element [2]. Figure 2.29 illustrates the structure of Is-limiter. A paralleled fuse with a higher

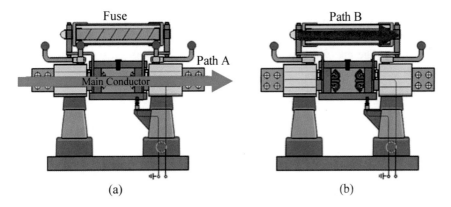

Fig. 2.29 Current path in Is-limiter [79]; **a** Before fault occurrence, **b** After fault occurrence

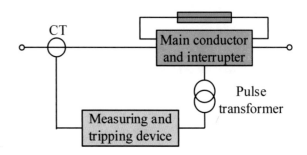

Fig. 2.30 Fault detection system of Is-limiter [79]

normal rated current conductor and detection system are involved in the Is-limiter. The conducting path provides an unrestricted way to flow load current. It allows using Is-limiter for the higher current applications; since the rated current in this device is determined by the main conductor, not fuse. The main role of the fuse is initiated after fault occurrence. As soon as the fault occurrence is identified by the fault detection unit, the main conductor is open in less than 1 ms and consequently, fault current is commutated to fuse. Finally, the fuse element melts and results in fault current interruption.

Figure 2.30 indicates the component of the fault detection system of Is-limiter [79]. The provided current by CT is transmitted to the measuring and tripping unit, which makes decision according to the measured current. In the case of detection of fault occurrence, it provides the tripping energy for pulse transformer. Finally, bus-bar potential is produced by converting the tripping pulse in pulse transformers. The setting of the tripping unit can be easily adopted considering system conditions. As a drawback, Is-limiter is not able to return in service after each operation. The outage time increases due to reset and replace the fuse after each operation.

The parallel connection of a reactor with Is-limiter adds the limitation function to this device, as shown in Fig. 2.31 [80]. In this case, fault current does not interrupt after melting fuse. It commutes to the reactor. The limited fault current is eventually interrupted by CB. This working principle provides more flexibility. After a fault is cleared, the load current can flow through the shunt reactor until servicing the Is-limiter. Until now, the maximum available rated current for Is-limiter has been 5 kA (at 750 V). It is also available up to 40.5 kV, but the maximum rated current is 2.5 kA [7]. The interrupting current is up to 210 kA. For higher rated current, the

Fig. 2.31 Parallel connection of Is-limiter with the reactor [80]

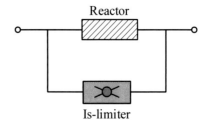

parallel installation of Is-limiters is suggested. According to the rated values, this device is applicable at LV and MV levels.

2.2.4 SSFCL

Advances in high-power superconductor switches such as silicon controlled rectifier (SCR), gate turn-off thyristor (GTO), super GTO (SGTO), insulated gate bipolar transistor (IGBT), integrated gate-commutated thyristor (IGCT), and emitter turn-off thyristor (ETO) have provided the possibility for developing power electronic-based FCL, which is known as SSFCL. Similar to SFCLs, different structures have been suggested for SSFCLs. Most of the designs work based on the current magnitude and emerge an additional impedance in the circuit during fault conditions. Indeed, they usually have a fault detection system. The modular design of power electronic converters can be used for SSFCL development, which is a desirable option. In order to deal with the high voltage and current ratings, the semiconductor blocks should be stacked or paralleled. This requirement can be handled through modular design. Additionally, no cryogenic system, the competitive cost in comparison with other FCL technologies, and compact design are the other benefits offered by SSFCL. However, thermal management, normal state voltage drop, conducting power loss, snubber circuit, and control units are the challenging issues related to SSFCL development. A large number of power electronic-based structures have been proposed for SSFCL in literature. In Ref. [3], SSFCL topologies were classified into three main groups: the series switch-type SSFCL, bridge-type SSFCL, and resonant-type SSFCL. Other classifications are also available in papers. In the following subsections, the basic concept of each group is discussed. However, the detailed explanations are ignored due to the high similarity of working principles.

2.2.4.1 Series Switch-Type SSFCL

Figure 2.32 illustrates the generic topology of series switch-type SSFCL. The parallel connection of bidirectional controlled semiconductor (S_{ss}), bypath switch (S_{bp}), current limiting impedance ($Z_{limiter}$), overvoltage protective devices (varistors), and snubber circuit fulfills the structure of the series switch-type SSFCL. SCR, ETO, GTO, SGTO, IGBT, and other controlled semiconductor switches can be applied in this concept. A snubber circuit is the essential component of the semiconductor switch. It suppresses the rate of rise of voltage across the switch (dv/dt) at the instant of turning-off to prevent the switch damage.

The series switch-type SSFCL also requires a fault detection system to take action. The fault detector is usually activated when the current exceeds the considered value (overcurrent schemes). Then the semiconductor switch is quickly turned off and current is transferred to the alternative path consisting of current limiting impedance. The mechanical switch is usually considered as the bypath switch in

Fig. 2.32 Series switch-type
SSFCL [3]

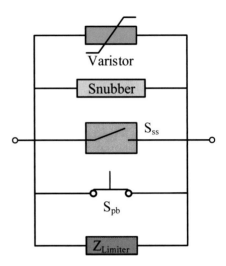

normal conditions. It has a lower conduction loss than the semiconductor switch. Besides, the waveform distortion is the lesser concern when load current flows through the mechanical switch instead of semiconductor one. The implementation of the mechanical switch is optional and some proposed structures work without it.

The purpose of the current limiting impedance is to reduce fault current to the permissible value and allows other protective devices to work safely. Resistive or inductive elements can be used for this branch, but the inductive element is more favored due to lower concerns about thermal management. This branch is optional similar to the mechanical switch. It can be eliminated in FCLs with interrupting capability (According to IEEE standard classification: Type B).

In order to protect the FCL against overvoltage, a high power varistor (usually, ZnO) should be implemented. When the semiconductor switch is turning off, varistor mitigates the provided voltage across the semiconductor switch by stored energy in line inductance through introducing an alternative current path. In addition to this task, varistors can be implemented across each semiconductor switch in series-connected switches to equalize voltage across them. The voltage across each switch may be different because of the turn-off or turn-on delays.

Based on the above-mentioned concept, different topologies have been proposed so far [3]. However, a few structures have been developed for high-capacity applications. EPRI and Silicon Power corporation have proposed, designed, and fabricated an FCL based on the series semiconductor switch. Figure 2.33 depicts the schematic of the suggested structure in the first phase of the project. As seen, there is no mechanical switch in this structure. It consists of SGTO switches, commutation circuit, varistor, and current limiting impedance. Antiparallel SGTOs and diodes are implemented in the main path. The commutation circuit is a series resonant circuit, which is activated during fault conditions. Lower switching loss of main SGTO switches and improving the current sharing problem are the main duties of this circuit.

Fig. 2.33 Proposed SSFCL by EPRI/Silicon Power (The first structure of the project) [3]

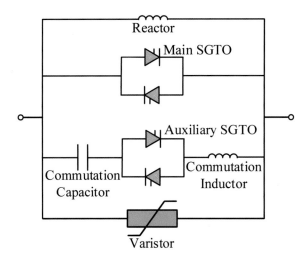

In the normal state, the main SGTO switches are turned on and provide a current path. The on-state voltage and conducting loss of used SGTOs are low. Therefore, SSFCL has a low voltage drop and power loss during this circumstance. As a fault happens in the power system, auxiliary SGTO switches are gated to provide the desired switching conditions for main SGTO switches. As soon as the main SGTO switches are turned off, the current is limited by current limiting impedance. It should be noted that the simpler method has been used in the later phase of the project as expressed in [81]. It uses the hard turn-off capability of the SGTO instead of SGTOs. The suggested SSFCL is modular. Standard building blocks (SBBs) can be connected together to achieve the desired voltage and current ratings. Modularity, immediate response, low recovery time, and compact design are the advantages offered by this SSFCL. More details about this project can be found in ref. [81].

2.2.4.2 Resonant-Type SSFCL

Similar to ResSFCL, the working principle of resonant-type SSFCL is based on the resonance phenomenon. Figures 2.34a, b illustrate the generic topologies of resonant-

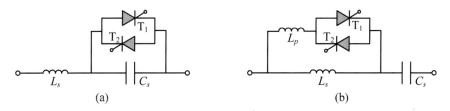

Fig. 2.34 Resonant-type SSFCL;

type SSFCL. There are also several evolved models based on this concept. In normal conditions, the semiconductor switches are in off-state and load current flows through the series LC circuit. The values of the inductor and capacitor are so considered that the resonance phenomenon occurs at line frequency (50 Hz or 60 Hz). Therefore, FCL introduces the low impedance and voltage drop in safe conditions. Once the occurrence of fault is detected by SSFCL, the semiconductor switches are activated. Then, fault current is limited either by bypassing a component like Fig. 2.35a or provide an additional component to the LC circuit like Fig. 2.35b. The consequence of both cases is to take the LC circuit out of the resonance state and a large impedance is finally introduced by SSFCL to suppress fault current.

The structure of this SSFCL is simple. In addition, semiconductor switches are only turned on under fault conditions, which is an astounding advantage. Since, the problems related to conducting loss and thermal management in the normal state are solved. However, SSFCL experiences a higher peak voltage and current stresses with the resonant circuit. The size and cost of the capacitor bank, ferroresonance of the inductor core, saturation of iron-core used in the inductor, and precise tuning of LC

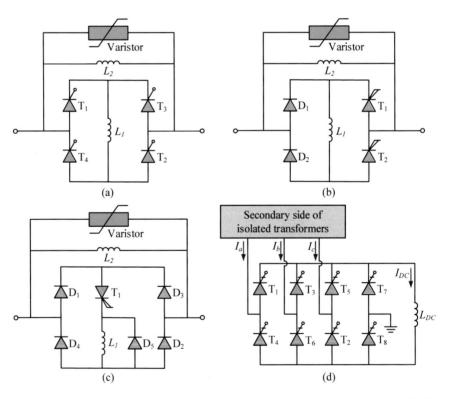

Fig. 2.35 Some proposed structure for bridge-type SSFCL [82–85]; **a** Thyristor bridge-type SSFCL with two reactors, **b** IGCT switch-based half-controlled bridge-type SSFCL, **c** Bridge-type SSFCL with single IGCT switch, **d** Transformer isolated GTO bridge-type SSFCL

circuit are challenging issues in resonant-based FCLs. Moreover, they may impose severe conditions on CB during current interruption due to produce high voltage across CB contacts.

2.2.4.3 Bridge-Type SSFCL

The operating principle of bridge-type SSFCL is the same as BSFCL. The implementation of a conventional DC reactor instead of the superconducting DC reactor (L_1) is the main difference between the two concepts. Lack of cryogenic system and lower initial and maintenance costs are the benefits of the DC reactor, but its power loss and dissipated heat are expected to be greater. Moreover, the use of controlled semiconductor switches instead of diodes has been investigated in this type of SSFCL. Figure 2.35 shows some proposed topologies reported in the literature for this type of SSFCL [82–85]. The controlled switches such as SCR, IGCT, GTO, and IGBT can be applied. They provide the flexibility in limiting percentage through pulse generation unit such that fault current can be even transferred to additional current branch consisting of AC reactor (L_2). AC reactors can endure fault current for a longer time and buy more time for the operation of other protective devices. This feature strictly depends on the capacity of the cooling system of FCL in the structure with this reactor. FCL may not able to limit fault current after a few cycles due to dramatic increase in temperature of semiconductor switches.

However, the controlled-based bridge-type SSFCLs require fault detector unlike diode bridge-type-based SSFCL, which operates spontaneously. The saturation problem in the DC reactor may also happen. In the case of using an additional reactor branch, the overvoltage protective devices are also required to alleviate the provided voltage across the FCL terminals at the instant of current interruption by CB. Additionally, more attention should be paid about the concerns including reliability, waveform distortion, switching loss, and thermal management.

Chapter 3
FCL Applications

Mohammad Reza Barzegar-Bafrooei, Jamal Dehghani-Ashkezari, Asghar Akbari Foroud, and Hassan Haes Alhelou

The power system has generally three main sections including power generation, transmission systems, and distribution systems. The high short circuit level will threaten all three parts of the future power system. Hence, all sections of the power system can be equipped with FCL in terms of short circuit level. In order to achieve the benefits offered by FCL in the power system, the optimal placement of FCL is required. The optimization problem with several main objectives such as reliability and cost is defined to determine the optimum number, impedance, and locations of FCL [86]. Figure 3.1 shows some candidate locations of FCL installation [2, 12, 21]. Table 3.1 lists these locations related to specified points in this figure. Furthermore, FCL can be used in ship propulsion systems and HVDC systems [6, 87]. Regarding published reports and papers related to the live-grid installation of FCL, bus bar coupling and incoming feeder are the potential locations for FCL installation. As known, to reduce power loss and voltage drop, transmission systems have a higher voltage than distribution systems and generation units. Although the development of FCL for all voltage levels is probably viable in the future, it is expected that FCLs are mainly applied at the MV level due to the following reasons:

M. R. Barzegar-Bafrooei (✉)
Department of Electrical Engineering, Ardakan University, Ardakan, Iran
e-mail: m.barzegar@ardakan.ac.ir

J. Dehghani-Ashkezari
Yazd Electrical Distribution Company, 8916794637 Yazd, Iran

A. Akbari Foroud
Electrical, Computer Engineering Faculty, Semnan University, Semnan, Iran
e-mail: aakbari@semnan.ac.ir

H. Haes Alhelou
The Department of Electrical and Computer Engineering, Isfahan University of Technology, Isfahan, Iran

© The Author(s), under exclusive license to Springer Nature Singapore Pte Ltd. 2022
M. R. Barzegar-Bafrooei et al. (eds.), *Fault Current Limiters*,
SpringerBriefs in Applied Sciences and Technology,
https://doi.org/10.1007/978-981-16-6651-3_3

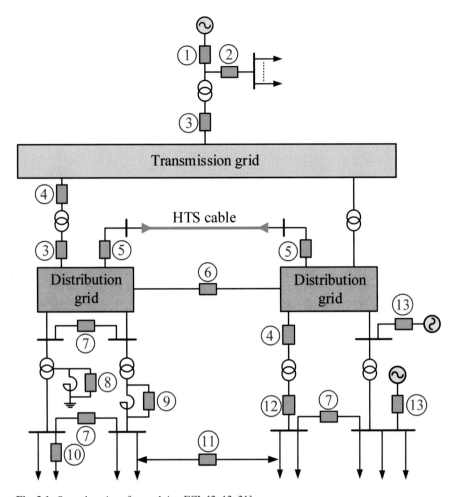

Fig. 3.1 Some locations for applying FCL [2, 12, 21]

- In HV applications, the utilization of reliable FCL is vital because of the large amount of transferred power. This subject may be a challenge for manufacturers or developers to convince electric companies for applying FCL in their power systems;
- The power system reliability generally reduces by an increasing number of power system equipment;
- Most DGs are connected to distribution systems. Therefore, the management of fault current in the distribution level is essential.

In the next subsections, some potential locations for the application of FCL are explained in detail.

Table 3.1 Description of specified locations in Fig. 3.4

Location Number	Description
1	Generator feeder
2	Power station auxiliary system
3	The secondary side of the transformer
4	The primary side of the transformer
5	The combination of FCL with other HTS devices
6	Network coupling
7	Busbar coupling
8	Shunting with neutral grounding impedance
9	Shunting with the current limiting reactor
10	Outgoing feeder
11	Loop-closing ring circuit
12	Incoming feeder
13	Coupling the local generation units or DGs

3.1 Busbar Coupling

Distribution systems are sometimes fed through parallel transformers for redundancy purposes. However, the parallel connection of transformers may considerably increase the total short circuit current due to the contribution of the sub-system. Even, the parallel connection may be avoided in extreme cases. The installation of FCL in busbar coupling eliminates the concerns related to short circuit current. Furthermore, the voltage level of the unfaulted bus during fault conditions is less severely affected and voltage sags and flickers decrease. During normal conditions, the operation of the parallel transformers results in lower power losses and voltage drops due to lower total impedance. The voltage distortions produced by non-linear loads decrease. The parallel connection provides higher system availability and allows connecting the larger loads in the sub-system. Additionally, the extra capacity of each bus can be used by another bus, and subsequently transformers experience better utilization. It should be noted that FCL also allows connecting busbar in the transmission substations. Splitting bus in high voltage is a common way to reduce fault current. A reliable FCL can be used instead of this technique.

3.2 Incoming Feeder

FCL in the incoming feeder can reduce the contribution of feeder in total fault current in the sub-system. It mitigates the imposed stresses on the transformer and CB and increases their useful lifetime. The upstream high voltage also experiences the lower voltage sag in the event of a fault. In the newly planned systems or replacement program, the transformers with lower short circuit impedance can be chosen when FCL is installed in the incoming feeder. This strategy reduces power loss and improves voltage regulation.

3.3   Outgoing Feeder

The contribution of the upstream system in the fault current of a particular feeder can be reduced by the installation of FCL in the outgoing feeder. The capacity of applied FCL in this location is lower than in other locations. Thus, the FCL is smaller and cheaper. It should be noted that the installation of FCL in the outgoing feeder should be avoided when a large number of outgoing feeders encounter with high short circuit level. Generally, in the outgoing feeders that replacement of equipment is difficult such as old underground cables utilization, the application of FCL is an appropriate option.

3.4   Generator Feeder

Synchronous generators and, in general, any rotating machines significantly contribute to the fault currents. It is obvious that the added generation units based on the synchronous generator to meet the demand for electric energy lead to an increase in fault current. Coupling an FCL in the generator feeder is an effective approach to alleviate fault current. Furthermore, it does not need to upgrade the power system equipment in the old substation. In the new project, the equipment with a lower short circuit current rating can be used. It should be noted that the reliability of FCL is vital in this location. In case of failure in the operation of FCL, generation losses may threaten the stability of the power system.

3.5   DG Application

The high penetration of DGs in the future power system is inevitable. Power electronic (PE)-based DGs usually have a low contribution in fault current due to applied fully rated converters in them. PE switches have a narrow overloading capability. However, other types of DGs would increase the short circuit level in distribution systems. Figure 3.2 shows an example of the application of FCL in the coupling of DG in an existing substation. The substation may face the potential fault current as of the large capacity DG or several DGs are connected. One expensive way to overcome this challenge is that DGs are connected to the high voltage grid via the transformer. The

Fig. 3.2 Application of FCL in Coupling of DG

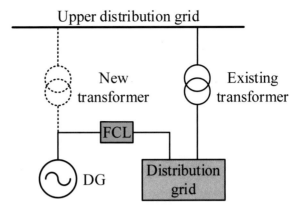

method can be avoided by implementing an FCL between DG and the distribution grid.

3.6 Power Plant Auxiliary System

The required electric energy of electrical appliances in a power plant consisting of fire/safety systems, offices, electric motors and pumps, control systems, and so on are served by power plant auxiliary system. The main generator of power plant usually feeds the power plant auxiliaries. However, the main concern in this system is the enormous fault current due to the vicinity of the main generator. The use of FCL in this location can reduce the fault current. Figure 3.3 illustrates the typical structure of the power system auxiliaries equipped with FCL. Arc faults in power plant auxiliaries have lower energy in the presence of FCL. It leads to lower requirements

Fig. 3.3 Application of FCL in power plant auxiliary system

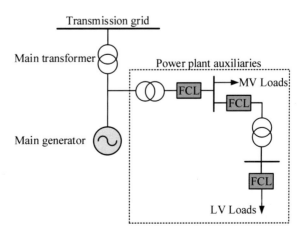

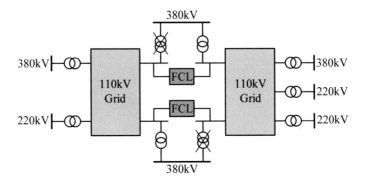

Fig. 3.4 Two subgrids coupled with FCL

for additional protection measures against these arcs. FCL in the location allows using lower size components especially in MV switchgear, which is cost-effective. Even, the possibility of using multicore cables with lower cross-section areas is provided.

3.7 Coupling of Subgrids

The distribution and transmission grids with lower transmission voltage levels are usually called subgrids. Through the step-down transformer, subgrids are powered by higher voltage systems. Subgrids are often separated to meet the acceptable short circuit current level or are connected through the additional transformer. It is possible to couple subgrids together by the installation of FCL. Higher power quality, lower power loss, and voltage drop during normal conditions are the benefits offered by the application of FCL in this location. The possibility of applying FCL in this location is assessed in the 110 kV system in Germany in [88]. Figure 3.4 shows the studied system.

3.8 Ship Propulsion Systems

All-electric ships need a significant amount of electrical energy owing to the utilization of electric propulsion systems instead of mechanical one. All loads in these ships use the new technology for connecting in the electric power system. 150 MW electric installation is a typical value for a navy ship. An example modern ship power system is shown in Fig. 3.5. The allowable voltage for ship power systems is typically low (between 5 and 15 kV). Therefore, it is expected that the short circuit level in the ship power system is high, even worse than the conventional power systems due to the local generation and consumption of power. Therefore, the management of fault current is essential in this system. The installation of FCL can cope with the high

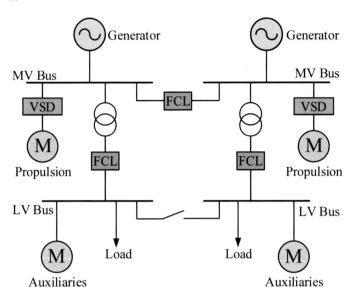

Fig. 3.5 Application of FCL in all-electric ships

fault current. It allows connecting the buses via a closed bus coupler without any concerns. Moreover, in the new ship, considering FCL may be a significant effect on the overall design. It should be noted that the main propulsion motors have no adverse effect on the fault current magnitude. Since, they are generally fed through the variable speed drive (VSD).

Chapter 4
Implementation Issues

Mohammad Reza Barzegar-Bafrooei, Jamal Dehghani-Ashkezari, Asghar Akbari Foroud, and Hassan Haes Alhelou

FCL as a new power system device may affect other aspects of the power system performance. The important prerequisite is to evaluate different aspects before the permanent installation. In critical cases, it is needed to consider some remedial actions. Many researches have been carried out to evaluate the interaction between FCL and power system since the emerging of FCL technology. The results are available in published papers and technical reports. Regarding the importance of the subject, the following issues should be taken into consideration when FCL is applied in the system:

4.1 Impact on the Performance of Protective Relays

The effect of FCL on the performance of conventional protective devices is considered as the main concern related to FCL application. The protective relays may make the mistake decision in the presence of FCL. In transmission systems, distance relays are mainly applied to protect the transmission line against different types of fault.

M. R. Barzegar-Bafrooei (✉)
Department of Electrical Engineering, Ardakan University, Ardakan, Iran
e-mail: m.barzegar@ardakan.ac.ir

J. Dehghani-Ashkezari
Yazd Electrical Distribution Company, 8916794637 Yazd, Iran

A. Akbari Foroud
Electrical, Computer Engineering Faculty, Semnan University, Semnan, Iran
e-mail: aakbari@semnan.ac.ir

H. Haes Alhelou
The Department of Electrical and Computer Engineering, Isfahan University of Technology, Isfahan, Iran

© The Author(s), under exclusive license to Springer Nature Singapore Pte Ltd. 2022   53
M. R. Barzegar-Bafrooei et al. (eds.), *Fault Current Limiters*,
SpringerBriefs in Applied Sciences and Technology,
https://doi.org/10.1007/978-981-16-6651-3_4

Conventional distance relays measure distance from relay location to faulty point by determining the apparent impedance. The input signals of the distance relay units are the voltage and current phasors at the fundamental frequency provided by VTs (or CVTs) and CTs. The apparent impedance seen by distance relay may considerably change in the presence of FCL. It leads to reduce accuracy and the sensitivity of the conventional distance relay. Studying the related papers in this subject shows that the parameters such as FCL type, FCL characteristics, FCL location, and type of fault affect the measured impedance by the relay [24, 25, 67, 89].

Furthermore, some FCL structures provide non-linear impedance when a fault occurs. Non-linear impedance distorts voltage and current waveforms and produces harmonics. Although modern protective devices can properly work under such conditions, the malfunction of old protective devices is not out of mind.

The working principle of the over-current relay (OCR) completely depends on the value of the current. The trip signal is sent to other protective devices when the current value becomes more than the setting of the relay. Among different OCR types, the relays with inverse time characteristics (such as IDMT) are common type in the distribution systems. The operation time of this type is inversely the function of current. It decreases with an increase in the magnitude of the current. The miss coordination between relays may happen when the system is equipped with FCL [90]. By reducing the magnitude of fault current, FCL has an adverse effect on the speed of fault clearing by inverse time characteristic-based OCR. Even, the relay may not detect fault occurrence in case that limited fault current is lower than the pick-up value of relay.

The use of FCL can also alter the coordination between recloser and fuses [91]. In contrast to recloser, fuses are usually installed near the load, meaning that they are far from the substation. When FCL does not exist in the system, the coordination between the fuse and recloser allows the loads to be in service. It is due to the proper setting between the setting of the recloser and the I-t curve of the fuse. However, FCL can deteriorate this coordination. According to the I-t curve of each fuse, the limited fault current may increase the melting time of fuse. In the critical case, the fuse does not melt. As the general outcome, the sooner operation of recloser and in turn service unavailability in healthy sections may happen in the presence of FCL.

It should be noted that the use of FCL can also help to keep the previous setting of protection schemes in some systems. For example, when a DG is connected to the distribution system, the short circuit level increases. Therefore, the revision in the coordination of protection schemes should be taken. The application of FCL in an optimum location not only can reduce fault current but also the restoration of the coordination of protection schemes is viable [92–93]. The optimum location can be achieved by the consideration of several factors such as FCL impedance and the contribution of DG in total fault current.

4.2 Impact on Wide-Area Protection

4.3 Impact on Interrupting Characteristics of CB

FCL is generally connected in series with CB. When a fault occurs, FCL quickly operates and reduces the fault current to the desired level. Finally, CB is activated and interrupts the limited current. The limited current satisfies one of the important parameters for the current interruption of CBs but not all. The interrupting characteristics of CB not only depend on fault current magnitude but also restrict by the recovery voltage produced across CB contacts. Recovery voltage is the appeared voltage across CB contacts after current interruptions (arc extinction) that can be characterized by transient recovery voltage (TRV) and power frequency recovery voltage. CB should be able to withstand the two mentioned voltages for successful operation. TRV is generated due to the response of the power system to the current interruption. When the instantaneous current becomes nearly zero during the current interruption, arc promptly extinguishes. Finally, the current flowing will be stopped after a few milliseconds. However, the voltage difference between the response of the power system placed at the upstream and downstream of CB (In some references: source side and load side) will provide TRV. It is clear that TRV depends on the circuit that should be interrupted. Hence, it is expected that FCL changes TRV waveshape and its important parameters due to adding a large impedance to the circuit. The peak of TRV and the rate of rise of recovery voltage (RRRV) are the two important parameters of TRV. Referring to performed studies, these parameters may experience better or severe conditions regarding FCL types [65–66, 94]. Therefore, further analysis based on actual conditions is needed before installation.

4.4 Impact on Voltage Sag Magnitude and Time Duration

Voltage sag or voltage dip is a common disturbance in the power system. It happens when the RMS voltage value decreases between 10 and 90% of the rated equipment voltage [95]. Voltage sag is mainly due to rapid changes in loads such as the overload and starting current of electric motors and short circuit faults. It has a short time duration, ranging from one-half cycle to 1 min. The shorter and longer phenomena are known as transient disturbance and undervoltage conditions, respectively. Voltage sag can be generally characterized by magnitude and time duration. It is clear that the use of FCL improves the voltage sage in terms of magnitude. However, voltage sag may last for a longer time due to a reduction in the sensitivity of protective relays in the presence of FCL [96].

4.5 Impact on System Reliability

The below items describe how an FCL can affect system reliability [5]:

- FCL reliability;
- FCL does not enter in current limiting mode when the power system needs it;
- FCL operates when no-fault occur in the power system (unwanted operations);
- FCL fails during limiting mode due to internal fault;
- Failure of additional auxiliary equipment installed with FCL;
- Maintenance necessities.

It should be also noted that the unwanted operation of FCL is mainly due to the impact of the power system on FCL. Starting current of motors, the transient current caused by single phase to ground fault in unground cable networks, and inrush current of transformer and capacitor bank switching are the possible cases that can cause the wrong operation of FCL.

References

1. A. Safaei, M. Zolfaghari, M. Gilvanejad, and G.B. Gharehpetian, A survey on fault current limiters: development and technical aspects, *International Journal of Electrical Power & Energy Systems*, Vol. 118, pp. 105729 (2020)
2. Technical Report. 1010760, "Survey of fault current limiter (FCL) technologies", EPRI, Palo Alto, CA, (2005)
3. A. Abramovitz, K.M. Smedley, Survey of Solid-State Fault Current Limiters. IEEE Transactions on Power Electronic **27**(6), 2770–2782 (2012)
4. CIGRE, WG A3. 10: "Fault Current Limiters", Report on the Activities of CIGRE WG 3.10, 2003
5. C. Li, B. Li, F. Guo, J. Geng, X. Zhang, and T. Coombs, "Studies on the active SISFCL and its impact on the distance protection of the EHV transmission line", *Protection and Control of Modern Power Systems*, Vol. 1, No. 1, 2016
6. M. Noe and M. Steurer, "High-temperature superconductor fault current limiters: concepts, applications, and development status", *Superconductor Science and Technology,* Vol. 20, pp. 15–29, 2007
7. ABB, Fault Current Limiters Is-Limiter and FC-Protector, 2017 [Online]. Available: https://library.e.abb.com/public/67ccc1b4ebff4e6f86ecc95fddf9c775/2562_FCL%20%20Is%20limiter%20FC%20Protector.pdf
8. IEEE std C37.302, 'IEEE guide for fault current limiter (FCL) Testing of FCLs rated above 1000 V AC', 2016
9. M.R. Barzegar-Bafrooei and A. Akbari Foroud, "Impact of fault resistance and fault distance on fault current reduction ratio of hybrid SFCL", *International Transactions on Electrical Energy Systems*, Wilely, Vol. 27, No. 11, pp. e2409, 2017
10. G. Didier, C.H. Bonnard, T. Lubin, J. Lévêque, Comparison between inductive and resistive SFCL in terms of current limitation and power system transient stability. Electric Power Systems Research **125**, 150–158 (2015)
11. CIGRE, WG A3. 16: "Fault current limiters", Report on the Activities of CIGRE WG A3.16, In Power Engineering Society General Meeting, 2006
12. M.R. Barzegar-Bafrooei, A. Akbari Foroud, Jamal Dehghani Ashkezari, and Mohsen Niasati, "On the advance of SFCL: a comprehensive review", *IET Generation, Transmission & Distribution*, Vol. 13, No. 17, pp. 3745–3759, 2019
13. B.W. Lee, J. Sim, K.B. Park et al., Practical application issues of superconducting fault current limiters for electric power systems. IEEE Trans. Appl. Supercond. **18**(2), 620–623 (2008)
14. L. Wang, J. Pengzan, W. Dada, "Summary of superconducting fault current limiter technology", *Frontiers in Computer Education.* Springer, Berlin Heidelberg **133**, 819–825 (2012)

© The Editor(s) (if applicable) and The Author(s), under exclusive license to Springer Nature Singapore Pte Ltd. 2022
M. R. Barzegar-Bafrooei et al. (eds.), *Fault Current Limiters*, SpringerBriefs in Applied Sciences and Technology, https://doi.org/10.1007/978-981-16-6651-3

15. X. Yuan, L. Kovalsky, K. Tekletsadik, J. Bock, F. Breuer, S. Elschner, Proof-of-concept test results of a superconducting fault current limiter for transmission-level applications. IEEE Trans. Appl. Supercond. **15**(2), 1982–1985 (2005)

16. S.M. Blair, C.D. Booth, A.N.D.G.M. Burt, Current-time characteristics of resistive superconducting fault current limiters. IEEE Trans. Appl. Supercond. **22**(2), 5600205 (2012)

17. H. Shimizu, Y. Yokomizu, T. Matsumura, N. Murayama, Proposal of flux flow resistance type fault current limiter using Bi2223 high TC superconducting bulk. IEEE Trans. Appl. Supercond. **12**(1), 876–879 (2002)

18. H. Shimizu, Y. Yokomizu, A study on required volume of superconducting element for flux flow resistance type fault current limiter. IEEE Trans. Appl. Supercond. **13**(2), 2052–2055 (2003)

19. T. Onishi, K. Sasaki, R. Akimoto, Investigation on reduction of required superconductor volume in a resistive fault current limiter with Bi2223 bulk superconductor. IEEE Trans. Appl. Supercond. **13**(2), 2100–2103 (2003)

20. A. Henning, M. Kurrat, Thermal–electric simulations of coated conductors with a variable conductivity of the buffer layer. IEEE Trans. Appl. Supercond. **17**(2), 3443–3446 (2007)

21. O.-B. Hyun, Brief review of the field test and application of a superconducting fault current limiter. Progress in Superconductivity and Cryogenics **19**(4), 1–11 (2017)

22. A. Hobl, W. Goldacker, B. Dutoit, L. Martini, A. Petermann, P. Tixador, Design and production of the ECCOFLOW resistive fault current limiter. IEEE Trans. Appl. Supercond. **23**(3), 5601804 (2013)

23. A. Colmenar-Santos, J.M. Pecharromán-Lázaro, C.d.P. Rodríguez, and E. Collado-Fernández, "Performance analysis of a superconducting fault current limiter in a power distribution substation", *Electric Power Systems Research*, Vol. 136, pp. 89–99, 2016.

24. S.R. Lee, J.J. Lee, J. Yoon, Y.W. Kang, J. Hur, Protection scheme of a 154-kV SFCL test transmission line at the KEPCO power testing center. IEEE Trans. Appl. Supercond. **27**(4), 1–5 (2017)

25. M.R. Barzegar-Bafrooei, A. Akbari Foroud, "Performance evaluation of distance relay in the presence of hybrid SFCL", *IET Science, Measurement &Technology*, Vol. 12, No. 5, pp. 581–593, 2018.

26. B.W. Lee, K.B. Park, J. Sim, I.S. Oh, H.G. Lee, H.R. Kim, O.B. Hyun, Design and experiments of novel hybrid type superconducting fault current limiters. IEEE Trans. Appl. Supercond. **18**(2), 624–627 (2008)

27. J.S. Kim, S.H. Lim, J.C. Kim, Comparative analysis on current limiting characteristics of hybrid Superconducting Fault Current Limiters (SFCLs) with first half cycle limiting and non-limiting operations. *Journal of Electrical Engineering and* Technology **7**(5), 659–663 (2012)

28. S.H. Lim, H.J. Ahn, C. Park, Study on fault current limiting characteristics of an SFCL using magnetic coupling of two coils with mechanical switch driven by electromagnetic repulsion force. IEEE Trans. Appl. Supercond. **24**(3), 5600704 (2014)

29. M. Ebrahimpour, B. Vahidi, S.H. Hosseinian, A hybrid superconducting fault current controller for DG networks and microgrids. IEEE Trans. Appl. Supercond. **23**(5), 5604306 (2013)

30. W. Paul, T. Baumann, J. Rhyner, Test of 100kw high-Tc superconducting fault current limiter. IEEE Trans. Appl. Supercond. **5**(2), 1059–1062 (1995)

31. D. Dersch, "Inductive current limitation device for an alternating current using the superconductivity superconductor", European Patent EP 0 353 449, 1994.

32. J. Kozak, M. Majka, S. Kozak, T. Janowski, Design and tests of coreless inductive superconducting fault current limiter. IEEE Trans. Appl. Supercond. **22**(3), 5601804–5601804 (2012)

33. O. Naeckel, "Development of an air coil superconducting fault current limiter," KIT Scientific Publishing, 2016.

34. J. Kozak, M. Majka, S. Kozak, Experimental results of a 15 kV, 140 A superconducting fault current limiter. IEEE Trans. Appl. Supercond. **27**(4), 5600504 (2017)

35. C. Schacherer, A. Bauer, S. Elschner, W. Goldacker, H.P. Kraemer, A. Kudymow, O. Nackel, S. Strauss, and V.M. Zermeno, "Smartcoil-concept of a full scale demonstrator of a shielded core type superconducting fault current limiter", *IEEE Transactions on Applied Superconductivity*, Vol. 27, No. 4, 2017.

36. Hayakawa, S. Chigusa, N. Kashima, S. Nagaya, and H. Okubo, "Feasibility study on superconducting fault current limiting transformer (SFCLT)", *Cryogenics*, Vol. 40, No. 4–5, pp.325–331, 2000.

37. S. Hellmann, M. Abplanalp, L. Hofstetter, M. Noe, Manufacturing of a 1-MVA-class superconducting fault current limiting transformer with recovery-under-load capabilities. IEEE Trans. Appl. Supercond. **27**(4), 5500305 (2017)

38. IEC60076–5: "Power transformers part 5: ability to withstand short circuit", 2000.

39. H. Kojima, M. Kotari, T. Kito, N. Hayakawa, M. Hanai, H. Okubo, Current limiting and recovery characteristics of 2 MVA class superconducting fault current limiting transformer (SFCLT). IEEE Trans. Appl. Supercond. **21**(3), 1401–1404 (2011)

40. M. Iwakuma, K. Funaki, H. Kanetaka, K. Tasaki, M. Takeo, K. Yamafuji, Quench analysis in a 72 kVA superconducting four-winding power transformer. Cryogenics **29**(11), 1055–1062 (1989)

41. H. Yamaguchi, T. Kataoka, K. Yaguchi, S. Fujita, K. Yoshikawa, K. Kaiho, Characteristics analysis of transformer type superconducting fault current limiter. IEEE Trans. Appl. Supercond. **14**(2), 815–818 (2004)

42. H. Yamaguchi, T. Kataoka, Effect of magnetic saturation on the current limiting characteristics of transformer type superconducting fault current limiter. IEEE Trans. Appl. Supercond. **16**(2), 691–694 (2006)

43. G. Wojtasiewicz, T. Janowski, S. Kozak, J. Kozak, M. Majka, B. Kondratowicz-Kucewicz, Experimental investigation of a model of a transformer-type superconducting fault current limiter with a superconducting coil made of a 2G HTS tape. IEEE Trans. Appl. Supercond. **24**(3), 5601005 (2014)

44. T.H. Han, S.H. Lim, Comparative study on current limiting characteristics of transformer type SFCL with common connection point between two secondary windings. IEEE Trans. Appl. Supercond. **28**(3), 5601705 (2018)

45. T.H. Han, S.H. Lim, Magnetizing characteristics of transformer type SFCL with additional secondary winding due to its winding direction. IEEE Trans. Appl. Supercond. **28**(4), 5602605 (2018)

46. T. Matsumura, T. Uchii, Y. Yokomizu, Development of flux-lock type fault current limiter with high Tc superconducting element. IEEE Trans. Appl. Supercond. **7**(2), 1001–1004 (1997)

47. S.H. Lim, H.S. Choi, D.C. Chung, S.C. Ko, B.S. Han, Impedance variation of a flux-lock type SFCL dependent on winding direction between coil 1 and coil 2. IEEE Trans. Appl. Supercond. **15**(2), 2039–2042 (2005)

48. S.H. Lim, H.G. Kang, H.S. Choi, S.R. Lee, B.S. Han, Current limiting characteristics of flux-lock type high-Tc superconducting fault current limiter with control circuit for magnetic field. IEEE Trans. Appl. Supercond. **13**(2), 2056–2059 (2003)

49. S.H. Lim, H.S. Choi, B.S. Han, Operational characteristics of a flux-lock-type high-Tc superconducting fault current limiter with a tap changer. IEEE Trans. Appl. Supercond. **14**(1), 82–86 (2004)

50. S.H. Lim, Operational characteristics of flux-lock type SFCL with the series connection of two coils. IEEE Trans. Appl. Supercond. **17**(2), 1895–1898 (2007)

51. M. Badakhshan, S.M. Mousavi, S,M, "Flux-lock type of superconducting fault current limiters: A comprehensive review", *Physica C: Superconductivity and its Applications,* Vol. 547, pp.51–54, 2018.

52. B. P. Raju, K. C. Parton, and T. C. Bartram, "A current limiting device using superconducting D.C. bias applications and prospects", *IEEE Transactions on Power Apparatus and Systems*, vol. APS-101, No. 9, pp. 3173–3177, 1982.

53. Y. Xin, W.Z. Gong, X.Y. Niu et al., Manufacturing and test of a 35 kV/90 MVA saturated iron-core type superconductive fault current limiter for live-grid operation. IEEE Trans. Appl. Supercond. **19**(3), 1934–1937 (2009)

54. A. Abramovitz, K. Ma Smedley, F. De La Rosa, and F. Moriconi, "Prototyping and testing of a 15 kV/1.2 kA saturable core HTS fault current limiter", *IEEE Transactions on Power Delivery*, Vol. 28, No. 3, pp. 1271–1279, 2013.
55. Y. Xin, W.Z. Gong, Y.W. Sun et al., Factory and field tests of a 220 kV/300 MVA statured iron-core superconducting fault current limiter. IEEE Trans. Appl. Supercond. **23**(3), 5602305 (2013)
56. C. Liang, C. Li, P. Zhang, M. Song, T. Ma, T. Zhou, Z. Ge, Winding technology and experimental study on 500 kV superconductive fault current limiter. IEEE Trans. Appl. Supercond. **28**(3), 5601105 (2018)
57. J.B. Cui, B. Shu, B. Tian, Y.W. Sun, L.Z. Wang, Y.Q. Gao, L. Liu, Z.Q. Wei, L.F. Zhang, X.H. Zhu, Q. Li, H. Hong, J.B. Cao, W.Z. Gong, Y. Xin, Safety considerations in the design, fabrication, testing, and operation of the dc bias coil of a saturated iron-core superconducting fault current limiter. IEEE Trans. Appl. Supercond. **23**(3), 5600704 (2013)
58. Y. Xin, W.Z. Gong, H. Hong, X.Y. Niu, J.Y. Zhang, A.R. Ren, B. Tian, Saturated iron-core superconductive fault current limiter developed at Innopower. AIP Conference Proceedings: Advances in Cryogenic Engineering: Transactions of the Cryogenic Engineering Conference-CEC **1573**, 1042–1048 (2014)
59. M.R. Barzegar-Bafrooei and A. Akbari Foroud, "The investigation of the current limitation behavior of saturated iron core SFCL under different fault resistance and fault distance", *27th Iranian Conference on Electrical Engineering* (ICEE2019), IEEE, Yazd, Iran, 2019.
60. J. Yuan, Y. Lei, C. Tian, B. Chen, Z. Yu, J. Yuan, J. Zhou, K. Yang, Performance investigation of a novel permanent magnet-biased fault-current limiter. IEEE Trans. Magn. **51**(11), 8403004 (2015)
61. J. Yuan, Y. Zhong, Y. Lei et al., A novel hybrid saturated core fault current limiter topology considering permanent magnet stability and performance. IEEE Trans. Magn. **53**(6), 8400304 (2017)
62. Y. Nikulshin, Y. Wolfus, A. Friedman, Y. Yeshurun, V. Rozenshtein, D. Landwer, U. Garbi, Saturated core fault current limiters in a live grid. IEEE Trans. Appl. Supercond. **26**(3), 5601504 (2016)
63. A. Pellecchia, D. Klaus, G. Masullo et al., Development of a saturated core fault current limiter with open magnetic cores and magnesium diboride saturating coils. IEEE Trans. Appl. Supercond. **27**(4), 5601007 (2017)
64. S. Quaia, F. Tosato, Reducing voltage sags through fault current limitation. IEEE Trans. Power Delivery **16**(1), 12–17 (2001)
65. M.R. Barzegar and M. Niasati, "Fusion TRV limiter to modify interrupting
66. characteristics of CBs with presence of resonance type SFCL", *30th International*
67. *Power System Conference (PSC)*, Tehran, Iran, pp. 92–98, 2015.
68. S.O. Faried, M. Elsamahy, Incorporating superconducting fault current limiters in the probabilistic evaluation of transient recovery voltage. IET Gener. Transm. Distrib. **5**(1), 101–107 (2011)
69. M.R. Barzegar-Bafrooei and A. Akbari Foroud, "Investigation of the performance of distance relay in the presence of saturated iron core SFCL and diode bridge type SFCL", *International Transactions on Electrical Energy Systems*, Wilely, Vol. 29, No. 2, pp. e2736, 2019.
70. J. Zhang, G. Zhiyuan, S. Naihao, W. Huaming, X. Liye, L. Liangzhen, Z. Zhenyu, Dynamic simulation and tests of a three-phase high T$_C$ superconducting fault current limiter. IEEE Trans. Appl. Supercond. **12**(1), 896–899 (2002)
71. S. Shimizu, O. Tsukamoto, T. Ishigohka, Y. Uriu, A. Ninomiya, Equivalent circuit and leakage reactances of superconducting 3-phase fault current limiter. IEEE Trans. Appl. Supercond. **3**(1), 578–581 (1993)
72. Y. Cai, Study on three-phase superconducting fault current limiter. IEEE Trans. Appl. Supercond. **20**(3), 1127–1130 (2010)
73. Y. Liu, M. Rong, Y. Wu, H. He, C. Niu, and H. Liu, "Numerical analysis of the pre–arcing liquid metal self–pinch effect for current-limiting applications", *Journal of Physics D: Applied Physics*, Vol. 46, No. 2, 2013.

74. H. He, Y. Wu, Z. Yang et al., Study of liquid metal fault current limiter for medium-voltage DC power systems. IEEE Transactions on Components, Packaging and Manufacturing Technology **8**(8), 1391–1400 (2017)

75. Y. Wu, H. He, M. Rong, A.B. Murphy, Y. Liu, C. Niu, X. Wu, The development of the arc in a liquid metal current limiter. IEEE Trans. Plasma Sci. **39**(11), 2864–2865 (2011)

76. K. Niayesh, J. Tepper, F. Konig, A Novel current limitation principle based on application of liquid metals. IEEE Trans. Compon. Packag. Technol. **29**(2), 303–309 (2009)

77. R. Strumpler, J. Skindh, J. Glatz-Reichenbach, J.H.W. Kuhlefelt, F. Perdoncin, Novel medium voltage fault current limiter based on polymer PTC resistors. IEEE Trans. Power Delivery **14**, 425–430 (1999)

78. Z. Guo, H. Sun, B. Bai, et al., "Experiment and simulation of PTC used in fault current limiter", *The 2010 China International Conference on Electricity Distribution*, pp. 1–9, 2010.

79. M.M. Arabshahi and K. Niayesh, "Design Of a matrix fault current limiter using low-power PTC resistors", *10th International Conference on Environment and Electrical Engineering (EEEIC)*, IEEE, 2011.

80. M. Steurer, K. Frhlich, W. Holaus, A novel hybrid current-limiting circuit breaker for medium voltage: Principle and test results. IEEE Trans. Power Delivery **18**(2), 460–467 (2003)

81. ABB, Is–Limiter, (Flyer) The World Fastest Limiting and Switching Device, 2012 [Online].

82. Available:https://library.e.abb.com/public/67b815bd4a0547eeb38efe2838cd3565/DS%202 302_0%20E%20Is_limiter%20Flyer.pdf

83. ABB, Is–Limiter, The World Fastest Limiting and Switching Device, 2015 [Online]. Available:

84. https://library.e.abb.com/public/e622c5385dcf05adc1257dce00341856/2493%20Is_Limi ter_GB_NewBranding.pdf

85. Technical Results 3002007174, "Solid-state fault current limiter development: design, build, and testing update of a 15.5-kv solid-state current limiter power stack", EPRI, Palo Alto, CA, 2015.

86. Lu, D. Jiang, and Z. Wu, "A new topology of fault-current limiter and its parameters optimization", *34th Annual Conference on Power Electronics Specialist (PESC03)*, IEEE, pp. 462–465, 2003.

87. W. Fei and Y. Zhang, "A novel IGCT-based half-controlled bridge type fault current limiter", *5th International Power Electronics and Motion Control Conference*, pp. 1–5, 2006.

88. W. Fei, Y. Zhang, and Q. Wang, "A novel bridge type FCL based on single controllable switch", *Power Electronics and Drive System*, pp. 113–116, 2007.

89. V. K. Sood and S. Alam, "3-phase fault current limiter for distribution systems", *International Conference on Power Electronics, Drive and Energy Systems*, pp. 1–6, 2006.

90. P. Yu, B. Venkatesh, A. Yazdani, B.N. Singh, Optimal location and sizing of fault current limiters in mesh networks using iterative mixed integer nonlinear programming. IEEE Trans. Power Syst. **31**(6), 477–4783 (2016)

91. A. Heidary, H. Radmanesh, K. Rouzbehi, J. Pou, A DC-reactor-based solid-state fault current limiter for HVd applications. IEEE Trans. Power Delivery **34**(2), 720–728 (2019)

92. C. Neumann, "Superconducting fault current limiter (SFCL) in the medium and high voltage grid", *IEEE Power Engineering Society General Meeting*, pp. 1–6, 2006.

93. M.R. Barzegar-Bafrooei and A. Akbari Foroud, "Studying an effective method to mitigate the adverse impacts of SFCL on transmission line distance protection", *IET Generation, Transmission & Distribution*, Vol. 13, No. 17, pp. 3823–3835, 2019.

94. R.M. Chabanloo, H.A. Abyaneh, A. Agheli, H. Rastegar, Overcurrent relays coordination considering transient behaviour of fault current limiter and distributed generation in distribution power network. IET Gener. Transm. Distrib. **5**(9), 903–911 (2011)

95. M.H. Kim, S.H. Lim, J.F. Moon, J.C. Kim, Method of recloser-fuse coordination in a power distribution system with superconducting fault current limiter. IEEE Trans. Appl. Supercond. **20**(3), 1164–1167 (2010)

96. W. El-Khattam W and T.S. Sidhu, "Restoration of directional overcurrent relay coordination in distributed generation systems utilizing fault current limiter", *IEEE Transactions on Power Delivery*, Vol. 23, No. 2, pp. 576–585, 2008.

97. A. Elmitwally, E. Gouda, S. Eladawy, Restoring recloser-fuse coordination by optimal fault current limiters planning in DG-integrated distribution systems. Int. J. Electr. Power Energy Syst. **77**, 9–18 (2016)

98. Q. Li, H. Liu, J. Lou, L. Zou, Impact research of inductive FCL on the rate of rise of recovery voltage with circuit breakers. IEEE Trans. Power Delivery **23**(4), 1978–1985 (2008)

99. P. Caramia, G. Carpinelli, P. Verde, *Power Quality Indices in Liberalized Markets* (Wiley, U.K., Chichester, 2009)

100. J.F. Moon, J.S. Kim, Voltage sag analysis in loop power distribution system with SFCL. IEEE Trans. Appl. Supercond. **23**(3), 5601504 (2013)

Printed in the United States
by Baker & Taylor Publisher Services